I0100409

The ISO 14001:2015 Implementation Handbook

The ISO 14001:2015 Implementation Handbook

Using the Process Approach to Build an Environmental Management System

Milton P. Dentch

ASQ Quality Press
Milwaukee, Wisconsin

American Society for Quality, Quality Press, Milwaukee 53203
© 2016 by ASQ
All rights reserved.
Printed in the United States of America
26 25 24 23 22   LS   8 7 6 5 4

Library of Congress Cataloging-in-Publication Data
Names: Dentch, Milton P., 1942– author.
Title: The ISO 14001:2015 implementation handbook : using the process
    approach to build an environmental management system / Milton P. Dentch.
Description: Milwaukee, Wisconsin : ASQ Quality Press, [2016] | Includes
    bibliographical references and index.
Identifiers: LCCN 2016006564 | ISBN 9780873899291 (hardcover : alk. paper)
Subjects: LCSH: Factory and trade waste—Handbooks, manuals, etc. |
    Environmental protection—Handbooks, manuals, etc. | ISO 14001
    Standard—Handbooks, manuals, etc.
Classification: LCC TD897.5 .D45 2016 | DDC 658.4/083—dc23
LC record available at http://lccn.loc.gov/2016006564

No part of this book may be reproduced in any form or by any means, electronic, mechanical, photocopying, recording, or otherwise, without the prior written permission of the publisher.

ASQ advances individual, organizational, and community excellence worldwide through learning, quality improvement, and knowledge exchange.

Bookstores, wholesalers, schools, libraries, and organizations: Quality Press books are available at quantity discounts with bulk purchases for business, trade, or educational uses. For more information, please contact Quality Press at 800-248-1946 or books@asq.org.

To place orders or to browse the full selection of Quality Press titles, visit our website at: http://www.asq.org/quality-press.

Quality Press
600 N. Plankinton Ave.
Milwaukee, WI 53203-2914
Email: books@asq.org

ASQ   Excellence Through Quality™

Table of Contents

List of Figures and Tables

Preface

I started my professional career in the paper industry in the early 1960s. The company I worked for, Rice Barton of Worcester, Massachusetts, produced machinery that manufactured pulp and paper. In my early twenties, I visited paper mills all over the United States and Canada. On my first visit to a paper mill, when I commented to my traveling companion from Rice Barton regarding the pungent odor as we approached the paper mill in Madawaska, Maine, my colleague advised, "That's the smell of money, son—get used to it." On a similar trip a few years later to a paper mill in Alabama, I experienced the alleged improved odor of "sweet southern pine," and it actually was less offensive.

I loved my job at Rice Barton; paper mills and mill towns were an important part of Americana. The engineering work was exciting—and other than the smell from the pulp mill stacks, I wasn't really conscious of the impact the paper industry had on air and water quality for the areas adjacent to the mills. A turning point for the paper industry occurred in the late 1960s when the cover of *Life* magazine was an aerial photograph of Lake Erie showing the runoff from the pulp mill at the Hammermill Paper Company in Erie, Pennsylvania (see Figure P.1). The mill discharged the effluent directly into the lake. The waste material was lighter than water, so the foamy material would rise to the surface and float many

Figure P.1 Aerial view of pulp waste from the Hammermill Paper Company in Erie, Pennsylvania, as it drains into Lake Erie, 1968. (Photo by Alfred Eisenstaedt/the LIFE Picture Collection/Getty Images)

yards offshore, contained by truck tires or similar barriers. With some frequency, the mill would skim off the flotsam and deposit the waste in a landfill. The toxic chemicals would remain in the lake, damaging all sorts of fish and wildlife.

While there were many other factors involved, the graphic *Life* magazine photograph illustrating the almost arrogant pollution of our lakes and rivers by industry helped spawn the Environmental Protection Agency (EPA) in 1970. That year also saw the publication of Rachel Carson's *Silent Spring*, the immensely popular book describing how the indiscriminate use of pesticides was poisoning birds and wildlife. Americans at large became aware of the ecology.

In 1969, I left the paper industry to work for the film and camera company Polaroid. The chemicals and materials used to produce instant film were often toxic; the processes included considerable wastewater and discharge of volatile air compounds. While Polaroid was always a responsible company, the required environmental control technology was in its infancy and safety trumped environmental concerns. "The solution to pollution is dilution" was the mantra, meaning keep adding air to chemical discharges to alleviate the released odor of volatile chemicals; a similar mind-set existed with management of wastewater.

The wake-up call for Polaroid was a front-page photograph in a Boston, Massachusetts, newspaper of several drums of hazardous waste that had washed up on the city's Revere Beach. Polaroid had contracted a company to dispose of the waste chemicals. We obviously did not conduct due diligence for this firm. Its disposal process was quite simple: take the drums a few miles out in Boston Harbor and dump them overboard. The workers would shoot bullet holes in the drums to ensure the partially full drums sank to the bottom of the sea. On this particular Saturday evening, the marksmen were not so accurate, and a half-dozen drums with Polaroid Corporation labels quite evident found their way to shore. Needless to say, Polaroid developed immensely improved controls for contractors over the next several years, and the company became a leader over the next decades in protecting the environment.

I worked as an engineer and manager for Polaroid for 27 years, holding positions with environmental responsibilities in several areas. While the company continued to reduce its environmental impact during my time there, there was something about the way companies like Polaroid managed their environmental programs that seemed less than ideal. We strove to obey the EPA and Massachusetts regulations. We sponsored "Earth Days," encouraging employees to reduce their environmental impact both at work and at home. But the individuals who managed the environmental programs were outside the mainstream manufacturing or engineering groups. The corporate environmental leaders were tasked by company management to "police" the manufacturing groups to ensure compliance with environmental regulations was met.

The corporate environmental leaders had associates in the various divisions, but these individuals were seen as caretakers of the environmental programs in their divisions, assumed to be mostly responsible for the environmental issues in their plant. The environmental associates maintained all the permits and other records associated with air, water, and waste controls. The "ownership" by the folks who produced the pollution was absent. When an associate left the division, the replacement sometimes had to scramble to locate the pertinent files.

Several years after leaving Polaroid, I became certified to provide audits for the International Organization for Standardization (ISO), first as a quality auditor,

Table P.1 The difference between a program and a system.

An environmental program:	An environmental management system requires:
• Can be dependent on individual knowledge	• Management oversight
• Can be reactive, with a compliance focus only	• A commitment to improve
• Can include inconsistent record keeping	• Formalized record keeping
• Can minimize employee involvement	• Employee involvement
• Can include a "silo" effect among managers	• Top management ownership and reviews
• Can be difficult to monitor	• Internal auditing

then later as an environmental auditor and EMS internal auditor trainer and consultant. Since 2001, I have audited the environmental management systems (EMSs) of over 100 companies to the International Standard ISO 14001. Observing what allowed some companies to implement a very successful environmental program, I discovered they weren't managing an *environmental program*; rather, these companies had created an *environmental management system*. The differences between a management system and a program are illustrated in Table P.1.

One of the goals of this book is to explain how an organization can use a management system to both control and improve its environmental performance. I provide guidance in building the EMS in support of the organization's operations—linking the management system to the requirements of ISO 14001, to support third-party certification to ISO 14001:2015. Included in the text are best practices as well as common pitfalls and weaknesses I've observed in various organizations. For those organizations already certified to ISO 14001:2004, I highlight the changes required to upgrade to the new International Standard.

In addition, included online are comprehensive check sheets to be used by internal auditors in auditing an EMS's conformance to ISO 14001:2015. Visit https://asqassets.widencollective.com/portals/56yurwh0/(H1507)Supplemental Files-TheISO140012015ImplementationHandbookUsingtheProcessApproach toBuildanEnvironmentalManagementSystem to access the check sheets.

Note: The contents of ISO 14001:2015 have been paraphrased in this book. Paraphrased text by its very nature can introduce differences in understanding and interpretation. This book should be used in conjunction with ASQ/ANSI/ISO 14001:2015 *Environmental management systems— Requirements with guidance for use.*

1

ISO 14001 History and Chronology

In 1968, the public outcry related to the contamination of Lake Erie by the discharge of waste from the pulp mill of the Hammermill Paper Company led to the eventual establishment of the United States Environmental Protection Agency (EPA) in 1970. On a worldwide basis, the leak of poisonous gas in 1984 from the pesticide plant in Bhopal, India, greatly amplified concerns on how chemical companies managed their environmental and safety operations, particularly in third-world countries. The discharge of the toxic gas, methyl isocyanate (MIC), from the Union Carbide India Limited (UCIL) plant in Bhopal, India, resulted in the deaths of thousands of employees and neighbors of the plant and caused serious injury to hundreds of thousands of others. Eventually, Union Carbide Corporation, parent company of UCIL, agreed to pay $470 million to the Indian government to be distributed to claimants as a settlement.

In response to the Bhopal tragedy, environmental management systems were established over the next several years:

- 1988: The American Chemistry Council (ACC) established Responsible Care to help member companies significantly improve their environmental performance and the health of the communities in which they operate

- 1992: The British Standards Institute (BSI) published the world's first environmental management systems standard: BS 7750, a standard for environmental control in the manufacturing and services sector

- 1993: The Eco-Management and Audit Scheme (EMAS) was established to allow industrial sector companies operating in the European Union to voluntarily participate in an environmental management scheme to demonstrate commitment to responsible environmental stewardship

- 1996: The International Organization for Standardization (ISO) created ISO 14001, a worldwide registration scheme that was compatible with BS 7750 and EMAS and consistent with the quality management system standard ISO 9001

ISO 14001 is a certifiable standard related to environmental management; it is similar to the quality management systems standard ISO 9001. The focus of ISO 14001 is to assist organizations in managing and improving their operations that affect the environment to comply with applicable laws and regulations. ISO 14001 is similar to ISO 9001 in that certification is performed by third-party organizations (registrars) rather than being awarded by ISO directly.

As of 2015, over 300,000 companies in 171 countries have been certified to ISO 14001. Since 1996, the "Big Three" automotive manufacturers—General Motors, Ford Motor Company, and Fiat Chrysler—along with their Japanese and German counterparts have mandated that their direct product suppliers (tier 1) achieve and maintain third-party certification to ISO 14001. The automotive manufacturers also require their assembly plants to maintain ISO 14001 certification. Additionally, many European and Japanese companies require their tier 1 suppliers to achieve ISO 14001 certification to demonstrate commitment to responsible environmental stewardship.

ISO has a goal to upgrade the management systems approximately every seven years. The first environmental standard, ISO 14001:1996 was reissued as ISO 14001:2004 in 2004 and had a delayed upgrade to ISO 14001:2015 due to the desire to harmonize the environmental standard with the quality management standard ISO 9001:2015.

2

The Environmental Management System as a Process

The internationally recognized standard for environmental management, ISO 14001 is built on the plan-do-check-act (PDCA) approach (see Figure 2.1). This is the operating principle of all ISO management system standards, including ISO 9001.

Put in the context of environmental management, the PDCA approach works as follows:

Plan: Top management establishes the scope and environmental policy of the EMS with consideration of the context of the organization's business model and interested parties. Environmental aspects and impacts with related compliance obligations are determined and analyzed to determine the risks related to maintaining the organization's environmental performance. Objectives are established to improve the environmental performance of the organization.

Do: Controls are implemented to ensure compliance with obligations. Environmental objectives and programs are initiated and implemented to improve the organization's environmental performance.

Check: The EMS is monitored and audited to measure performance against the organization's objectives and compliance obligations. The performance and results of the EMS are reported.

Act: Actions are initiated to correct deficiencies and improve the environmental performance as indicated by the monitoring and measurement of the EMS results. Resources and employee training are provided as appropriate to ensure improvement of the EMS.

While establishing the plans and actions to support an EMS, it is helpful to look at the EMS as a *process* with two desired outputs: compliance with applicable environmental regulations and improvement of the EMS. The organization's management provides the inputs to the EMS process: scope of activities (business model), the environmental policy, and regulated environmental activities (aspects).

The chart shown in Figure 2.2 represents the core processes of an EMS and is the starting point for building the EMS. The next step is to define the business model for the organization with linkage to related ISO 14001:2015 requirements.

For organizations looking to certify to ISO 14001 for the first time, I recommend a review of what is already in place in the business related to the core EMS process before attempting to conform to the ISO requirements. Unfortunately, there is

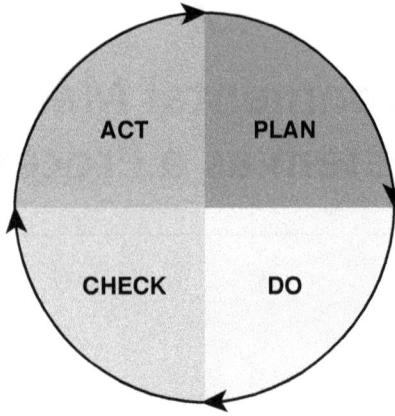

Figure 2.1   Plan-do-check-act cycle.

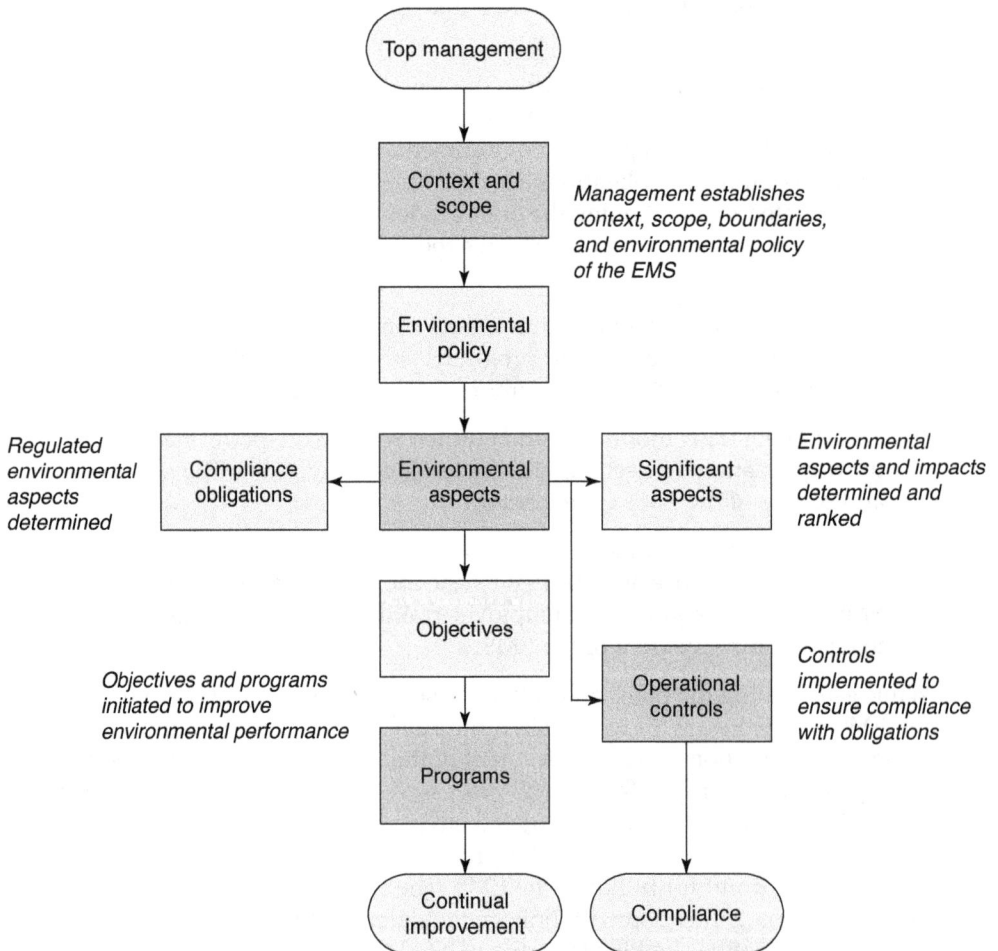

Top management

Context and scope

Management establishes context, scope, boundaries, and environmental policy of the EMS

Environmental policy

Regulated environmental aspects determined

Compliance obligations

Environmental aspects

Significant aspects

Environmental aspects and impacts determined and ranked

Objectives

Objectives and programs initiated to improve environmental performance

Operational controls

Controls implemented to ensure compliance with obligations

Programs

Continual improvement

Compliance

Figure 2.2   The core processes of an EMS.

quite a bit of "ISO-speak" in all International Standards, due to the need to cover organizations of various sizes and complexities, with a multitude of worldwide languages and interpretations. So, an organization with a high or low environmental impact can collect existing process controls and documentation in support of the EMS process model shown in Figure 2.2 before engaging in steps to conform to ISO 14001:2015. The goal should be to achieve and maintain compliance with the organization's environmental regulatory requirements, while improving its environmental performance. Becoming certified to ISO 14001 provides verification by a third party, certified by ISO.

For organizations currently certified to ISO 14001:2004, this book highlights and explains what's needed to satisfy ISO 14001:2015. The major changes from ISO 14001:2004 are summarized in the following section.

CHANGES FROM ISO 14001:2004

Organizations currently certified to ISO 14001:2004 will need to address the new (or expanded) requirements of ISO 14001:2015 with the following general groupings:

- Understanding the context of the organization and expectations of interested parties

- Integration of the EMS requirements into the organization's business processes

- Actions to address risks and opportunities

- Expanded top management commitment

- Expanded definition of operational controls

Context and Interested Parties

Past revisions of ISO 14001 required organizations to define the *scope*—the activities, processes, and buildings and property within their EMS. The organization's environmental policy included commitments to comply with applicable environmental regulations, reduce pollution, and continually improve its environmental performance. While many ISO 14001–certified organizations included initiatives such as replacement of toxic materials, recycle activities, and reduction of fossil fuel use, the majority of the environmental programs fostered by certified organizations were "reactive" in nature. The new clauses of ISO 14001:2015, Understanding the organization and its context and Understanding the needs and expectations of interested parties, challenge organizations to analyze their operations and environmental impact from a more holistic, proactive vantage point.

With regard to the context of the organization, depending on the organization's business model and environmental impact, there may be opportunities to have programs with a positive, proactive impact: air emissions related to climate change, improvement in soil quality of adjacent land or neighboring waterways, and supporting biodiversity (local flora-fauna) initiatives. The organization should be able to explain, within the context of its operations, what opportunities exist. A company producing machined parts or electronic components may have limited options; a large chemical plant, oil refinery, or paper mill may have many opportunities.

In a similar fashion, in response to expectations of interested parties, the organization should analyze how its activities and products may have an environmental impact on its customers, community, and neighbors. Examples could be end-of-life product disposal, voluntary labeling on products, reduction of unregulated materials, sustainable resources, and commitments to maintain adjacent community land or waterways. A key focus of ISO 14001:2015 is "life-cycle thinking," considering each stage of a product or service, from development to end of life. Organizations producing consumer products may have many opportunities (but also challenges) to have an impact on product disposal, while manufacturers of components for sale to industry may be limited in this regard. Organizations providing components for sale to industry can demonstrate their commitment to life-cycle thinking by reducing or eliminating the use of environmentally challenged materials and maximizing recycle initiatives. Consumer product manufacturers can support life-cycle thinking by moving to "zero-landfill" disposal via waste-to-energy disposal options (waste is burned to produce electricity). Chapter 4 discusses these requirements in more detail.

Integration of EMS Requirements into the Organization's Business Processes

Many ISO 14001–certified companies have integrated the EMS into their business planning and strategy. I have audited companies of all sizes where the environmental performance metrics are woven into the business plan; the key process indicators (KPIs) assigned to quality and business parameters include the environmental metrics of hazardous waste reduction, material recycle, and utility use. Quality-driven waste-reduction projects include improved environmental performance. Best-in-class organizations have established a business management system (BMS) incorporating their financial, quality, safety, and environmental systems into a cohesive operational model. I have also experienced ISO 14001–certified companies that operate with their EMS at arm's length from their business—just doing the minimum in environmental management to maintain certification. The ISO 14001:2015 standard requirements should nudge these companies into broadening their perspective on environmental performance. Chapter 9 discusses ways to integrate the EMS into the organization's business.

Actions to Address Risks and Opportunities

From my perspective, the requirement to provide risk analysis in the EMS activities is the key difference between ISO 14001:2015 and previous revisions. An identified method to analyze risk associated with threats and opportunities related to the organization's significant environmental aspects, compliance obligations, or other issues is now required. Results of the analysis should be used in establishing objectives and planning to mitigate the risks. Utilization of failure mode effects analysis (FMEA) as a quality tool could be applied. While organizations with an effective EMS certainly understand risks related to noncompliance, the new requirements of ISO 14001:2015 may have a positive effect on many organizations by requiring a more formalized process and subjecting the risk evaluation process to a third-party audit. Chapter 6 discusses this requirement in more detail with examples of application to environmental controls and legal obligations. It should be noted that ISO 14001:2015 does not have a requirement for "preventive action." The thought is that the entire EMS is *preventive* in nature and that the risk analysis approach is also preventive.

Top Management Commitment

While the previous revisions to ISO 14001 included commitment from management to support the EMS, ISO 14001:2015 amplifies this commitment. The ISO 14001:2015 standard does not use the title "Management Representative" as previous versions did. The organization can continue to use "Management Representative" as the title having certain responsibilities, but the intent of ISO 14001:2015 is to emphasize top management's responsibilities as more than delegating. My past experience with a small group of organizations was that management had delegated the environmental management coordination too far down the organizational chart. This was evident by way of the environmental management representative not attending the management review meetings to present the status of the EMS. Not a good sign. Management would justify this by citing the need to discuss financial or sensitive issues at the meeting, and the environmental coordinator should not be privy to such information. ISO 14001:2015 requirements strive to prevent the over-delegation of the EMS support and coordination.

Expanded Definition of Operational Controls

The intent of ISO 14001:2015 clause 8 closely matches the previous revisions of ISO 14001 related to operational controls; however, ISO 14001:2015 provides more specific requirements related to outsourced activities (contractors and suppliers), change control, design, and product life-cycle considerations.

When processes are provided by contractors and/or suppliers (outsourced), the environmental impacts need to be defined with appropriate controls established and implemented. Examples of outsourced processes include cleaning services, waste removal, landscaping, contracted maintenance, and construction.

The product design process should consider the impact of materials used in order to avoid environmentally challenged materials during use, delivery, and disposal at end of life. The design and life-cycle requirements of ISO 14001:2015 are mostly applicable to organizations manufacturing consumer products. Other organizations, such as discrete manufacturers, original equipment suppliers, and contract electronics manufacturers, are restricted by compliance obligations such as the RoHS (Restriction of Hazardous Substances) Directive and REACH (Registration, Evaluation, Authorisation and Restriction of Chemicals).

The organization needs to ensure controls are in place when changes occur in processes, resources, and equipment. This requirement was *implied* in ISO 14001:2004 clause 4.4.6, Operational control (and clause 4.3.1, Environmental aspects); however, ISO 14001:2015 now *explicitly* requires the organization to evaluate and adjust operational controls when new processes or equipment is added to the operation.

Chapter 8 provides additional detail on the expanded requirements in clause 8 of ISO 14001:2015.

Auditor Interpretations

The new or expanded ISO 14001:2015 requirements outlined here will be somewhat subjective for the third-party auditors to evaluate. When will an organization be judged nonconforming in addressing the context/interested parties, risk analysis, or top management commitment? If there are no examples of proactive initiatives related to reducing the organization's environmental impact, is that a nonconformance? My past experience in auditing to the ISO 14001 environmental standards (and ISO 9001 for quality) was that the auditor would not issue nonconformances when the organization did not meet its improvement objective or goal—provided the organization either documented the reason why the goal was missed or established actions to correct the situation. At a minimum, an organization is expected to have some form of "risk analysis" process related to the EMS. I would expect third-party auditors to follow this guideline when assessing performance against the new ISO 14001:2015 requirements. With regard to assessing top management commitment to the EMS, an experienced third-party auditor can detect when resources are not adequate to support the EMS—and will issue nonconformances as applicable.

Two other changes (but not new requirements) are the modification of the clause outlines and documentation formatting. These changes were made in ISO 14001:2015 to provide alignment with the formatting of ISO 9001:2015. Chapter 7 outlines these adjustments. Appendix A, "Correspondence: ISO 14001:2015 to ISO 14001:2004," provides clause-by-clause correspondence of the two standards.

Transitional Period of ISO 14001:2015

ISO 14001:2015 was published on September 15, 2015. Companies that are certified to ISO 14001:2004 will have three years to bring their EMS up to date with ISO 14001:2015. Eventually all certificates in accordance with ISO 14001:2004 will become invalid and will be withdrawn as of September 15, 2018.

It is usually more efficient for both the organization and the ISO registrar to conduct the upgrade audit to ISO 14001:2015 during the organization's three-year recertification audit; however, the upgrade can occur during the annual surveillance audit.

ISO 14001 BENEFITS

I have presented overviews of ISO 14001 to managers and executives at several large companies over the past several years. The agenda for the meetings typically included:

- Major elements of ISO 14001

- ISO 14001 common "gaps"

- ISO 14001 best practices

- Benefits of an EMS

- Role of senior management

- Steps toward third-party registration

- Questions and wrap-up

Note: Appendix B summarizes the key points from the presentation and may be useful for organizations looking to certify to ISO 14001:2015. A review of the benefits of ISO 14001 is included in this chapter before moving on to defining the requirements of ISO 14001:2015.

Most management teams want to understand the return on their investment in an EMS and why they should obtain third-party certification to ISO 14001. What are the benefits of an EMS? In many cases, the company is already "committed" to certifying to ISO 14001 by way of a requirement placed on them by customers, often from the automotive sector or firms from Europe or Asia. Depending on the organization's business model and environmental impact, potential benefits in certifying can be grouped into three categories: risk reduction, cost reduction, and enhanced competitive edge.

Risk Reduction

- Environmental legal liability

- Accidents and environmental damage

Cost Reduction

- Disposal costs

- Utility costs

- Permitting fees

Competitive Edge

- Improved corporate image

- Investment in long-term stability

- Improved relations with regulators

- Counter international market pressures

- Strategic investment now versus necessary expense later

- Demonstration of leadership

It can be difficult to place hard dollar amounts on the savings or cost avoidance in managing an organization's EMS. In relative terms, the cost of implementing an ISO 14001 system as a preventive action versus the cost of environmental noncompliance is illustrated in Table 2.1.

My "exponential" ranking may be soft in today's noncompliance penalties process, where regulatory fines can be in the hundreds of thousands of dollars. A local newspaper recently reported on a laboratory that was fined by the Massachusetts Department of Environmental Protection (MassDEP):

> An environmental testing company based in Marlborough will pay $100,000 to resolve allegations it failed to comply with hazardous waste laws, in violation of a previous court judgment.

Table 2.1   Relative costs of ISO 14001: prevention versus response to errors.

Cost of ISO 14001 as a prevention tool	Cost of ISO 14001 as a discovery tool	Cost of internal environmental errors	Cost of external environmental errors
• Training • Maintenance • Process improvement • New product design • Communication	• Inspections • Audits • Monitoring • Reporting • Calibration	• Loss of materials • Waste treatment and control • Loss of productive space • Loss of time	• Fines and plant closures • Remediation • Loss of customers • Reaction to unfriendly community
$1	$10	$100	$1000

The Massachusetts Attorney General's Office announced Thursday that Accutest Laboratories of New England, Inc. recently reached a settlement with the state after being accused of failing to make improvements at its facility as required by a previous court ruling.

State officials allege the company improperly stored hazardous waste for longer than permitted, did not properly label hazardous waste containers and violated other waste storage rules.

State environmental officials also allege the company's efforts to prevent emergencies at its analytical laboratory in Marlborough were inadequate, and that it should have provided more training for employees about handling hazardous waste.

"This company ignored the terms of a court judgment," Attorney General Maura Healey said in a prepared statement. "We expect defendants in all cases, including companies caught violating environmental laws that protect public health and natural resources, to take their compliance obligations seriously."

Accutest was ordered to pay a $350,000 fine in May 2014, settling prior allegations the company failed to obtain necessary permits to emit large quantities of hazardous pollutants into the air.

According to Healey's office, the company was accused of underreporting the level of hazardous waste it was generating to the state Department of Environmental Protection. It was ordered to apply for a new air permit and install emission control equipment to reduce emissions by 95 percent, according to Healey's office.

Inspectors from MassDEP visited Accutest's facility in Marlborough again in October 2014 and found that, while the company had taken "appropriate steps" to comply with air permitting requirements, the building still did not comply with hazardous waste laws, according to the state.

The company has since corrected the problems. It was also ordered to hire an environmental health and safety officer to ensure it is complying with environmental laws. The employee must remain on staff while the court order is in effect—a period of at least three years, according to Healey's office. (Haddadin 2015)

The irony of the citation is that the laboratory performed *environmental* testing services including organic and inorganic analysis of air, water, soil, and waste

Table 2.2   Environmental violations at laboratory as related to ISO 14001.

ISO 14001 clause	Violation
Operational controls	Improper labeling of hazardous waste containers and violation of other waste storage rules
Training	Employees did not receive necessary amount of training for handling hazardous waste
Compliance obligations	Failed to obtain necessary permits to emit large quantities of hazardous pollutants into the air
	Underreported the level of hazardous waste it was generating to the state Department of Environmental Protection
Emergency planning	Plan to prevent emergencies at its analytical laboratory was inadequate

characterization in support of federal and state environmental programs. The firm's website did not indicate certification to ISO 14001 (or ISO 9001).

The violations cited provide both an excellent segue to subsequent chapters of *The ISO 14001:2015 Implementation Handbook* and justification for becoming certified to ISO 14001. A properly implemented EMS requires controls and auditing for all the issues raised by MassDEP (see Table 2.2).

Additionally, an ISO 14001–certified company is required to perform internal and compliance-based self-audits. Either type of audit would have helped the laboratory address the issues before they became external violations. In similar cases, the EPA has mandated that the offending organization require certification by a third party, often a registrar providing ISO 14001 certification. I have audited several firms to ISO 14001 based on responses to violations cited by the EPA. Appendix D, "United States Environmental Protection Agency Enforcement Annual Results for Fiscal Year (FY) 2015," describes other notifications of violations raised during 2015.

Cost Reductions and ISO 14001

In addition to avoiding the costs of environmental noncompliance, organizations that establish a formal EMS and become certified to ISO 14001 can benefit by reduced disposal costs, utility costs, and permitting fees. Some possibilities I have encountered at various organizations are listed in Table 2.3.

Many organizations without ISO 14001 certification are already achieving the cost savings outlined by exercising good business practices and environmental management. Well-managed companies, when seeking ISO 14001 certification (often due to customer decree), are challenged to demonstrate improvements in their environmental performance, as they have already harvested the low-hanging fruit. In my experience auditing many companies, there are always opportunities for improvements, particularly in large manufacturing plants. Reduction of utility costs, energy savings, material substitution, and recycling programs are an ongoing challenge for all in industry. For smaller firms with minimal environmental impact, an experienced third-party auditor will understand their challenge and review other facets of the organization's environmental performance to assess performance.

While auditing a small, low environmental impact machine shop that was forced into ISO 14001 by a customer, I was informed by the plant manager: "Well,

Table 2.3   Cost benefits of ISO 14001.

Area	Possibilities
Disposal costs	• Maximize recycle of materials to both reduce landfill fees and reuse materials • Optimize segregation of waste to increase value of recycled materials • Invest in trash compactors • Establish composting field to reduce landfill load • Replace or reduce hazardous materials • Distill water from liquid waste to reuse water and reduce disposal cost • Partner with quality department to improve material yields to reduce landfill load • Maximize design opportunities to use environmentally compatible materials
Utility costs	• Partner with electricity supplier to relamp factory • Install motion detectors to reduce lighting • Invest in high-efficiency motors to reduce electrical use • Schedule production to leverage power factors in evening • Upgrade insulation and door openings to save energy • Monitor air compressor losses during non-production periods • Recycle wash water • Install water meter on cooling towers to lower cost of sewer charges from city • Optimize start-up and shutdown of machines • Establish employee energy savings team
Permitting fees	• Eliminate or reduce quantity of hazardous waste to lower waste category and fees • Reduce air emissions of volatile organic compounds by material substitution to avoid permit requirement

this ISO certification will cost us $10,000 this year. I told the guys—we better find some way to pay for it." And, they did. The shop put a push on separation of metal waste types. They found that by developing a process to remove the lubricating oil, the recycled material was more valuable to the recycling company. Also, by managing the quantity of recycled material to equal a full truckload, savings were generated in shipping. The savings continued into the next year as well.

3
ISO 14001:2015 Requirements

In the 2015 revisions, ISO formatted the management systems in sections for both quality (ISO 9001:2015) and environmental (ISO 14001:2015):

1. Scope
2. Normative references
3. Terms and definitions
4. Context of the organization
5. Leadership
6. Planning
7. Support
8. Operation
9. Performance evaluation
10. Improvement

1 SCOPE

This International Standard specifies the requirements of an EMS for organizations seeking to establish, implement, maintain, and continually improve a framework with the aim of managing its environmental responsibilities in a manner that contributes to the "environmental pillar" of sustainability. The intended outcomes of an EMS provide value for the environment, the organization, and its interested parties. Consistent with the organization's environmental policy, the intended outcomes of an EMS include enhancement of environmental performance, conformance to compliance obligations, and fulfillment of environmental objectives. This International Standard is applicable to any organization regardless of size, type, or nature and applies to the environmental aspects that the organization determines it can either control or influence considering a life-cycle perspective. It does not state specific environmental performance criteria, nor does it increase or change an organization's legal obligations.

This International Standard can be used in whole or in part to improve environmental management, but all the requirements are intended to be incorporated into an EMS and fulfilled, without exclusion, if an organization claims it complies with this International Standard.

2 NORMATIVE REFERENCES

No normative references are cited. This clause is included to maintain clause numbering alignment with other ISO management system standards.

3 TERMS AND DEFINITIONS

See Appendix C (refer to pp. 1–6 of ASQ/ANSI/ISO 14001:2015).

4 CLAUSES 4–10

Sections (clauses) 4–10 provide the requirements for certification to ISO 14001:2015.

Note: In the chapters following, the general requirements of each ISO 14001: 2015 clause are paraphrased in the opening box. "Needs to" and "should" indicate requirements. Where requirements in ISO 14001:2015 have changed from ISO 14001:2004, the issue is highlighted in the text in **bold**.

4

Clause 4: Context of the Organization

4.1 Understanding the organization and its context

The organization needs to determine the external and internal issues that are relevant to its purpose and that affect its ability to achieve the intended outcomes of its environmental management system.

4.2 Understanding the needs and expectations of interested parties

The organization needs to define the interested parties and their needs and expectations relevant to its environmental management system, including its compliance (legal and regulatory) obligations.

4.3 Determining the scope of the environmental management system

The organization needs to determine the boundaries and applicability of the environmental management system to establish its scope: the activities, organizational units, functions, and physical boundaries included in the organization's environmental management system.

4.4 Environmental management system

The organization needs to establish, implement, maintain, and continually improve its environmental management system, in accordance with the requirements of this International Standard.

In previous revisions of ISO 14001, the organization was required to define and document the scope of its EMS. There was some confusion as to what was meant by "scope." Could the organization exclude certain activities at the site? How about organizations that lease or rent the site? What was the leaseholder responsible for in environmental controls? In **ISO 14001:2015,** the context, scope, and boundaries of the organization's EMS need to be clearly defined to provide a baseline for control and monitoring. In general terms, the scope, boundaries, and context as related to an EMS can be explained as follows.

SCOPE

The scope is what the organization does, the products or services the organization provides. For organizations with a quality management system (QMS), the scope is what the organization states on its ISO 9001 certificate and advertising media. For example, a plastic manufacturer might have "Manufacturer of injection molded products" as its scope. This scope would apply for this organization's EMS, but more information is needed in the scope to describe the boundaries included in the organization's responsibilities and the context of its commitments.

BOUNDARIES

The scope needs to define the spatial and organizational boundaries to which the EMS will apply, especially if the organization is part of a larger organization at a given location. How many sites (building addresses) are under the scope? Who owns the property? If the organization does not own the site(s) but leases the properties from a landlord (leaseholder), then the responsibilities for the organization and the landlord need to be defined. What are the boundaries? Are there manufacturing or service groups located on the organization's site that are not in the scope of the EMS? The organization needs to define and clarify environmental aspects and impacts under its control.

The boundaries and site ownership should not be considered a form of "exclusion" or "outsourcing." In the case where the organization leases the property, the organization has responsibility to ensure its processes or activities do not have an adverse environmental impact on the leaseholder's property. While the leaseholder may hold the storm water or wastewater permits for the site, there should be clear definition as to the organization's responsibilities. Likewise, if the organization rents out portions of the site to another firm, the organization has responsibility to ensure the renter does not have an adverse impact on the organization's environmental performance. A best practice in these situations is to have *documented information*, such as a matrix, describing environmental aspects and leaseholder (or renter) responsibilities versus the organization's responsibilities.

ISO 14001:2015 defines documented information as "information required to be controlled and maintained by an organization and the medium on which it is contained." This change is described in Chapter 7.

CONTEXT OF THE ORGANIZATION

ISO 14001:2015 requires organizations to put a policy in place that promotes environmental protection specific to the context of their business. Previous versions of ISO 14001 placed environmental protection in a somewhat reactive mode, where proactive initiatives included recycle programs and pollution prevention directly related to the organization's manufacturing processes. With ISO 14001:2015, the organization is required to understand the important issues that can affect, either positively or negatively, the way it manages its environmental responsibilities.

Examples include:

- Reduction of air emissions related to climate change

- Use of sustainable resources

- Improvements in soil quality of adjacent land or neighboring waterways

- Support of biodiversity (local flora-fauna) initiatives

In addition to understanding the context of the organization, **ISO 14001:2015** requires the organization to understand the needs and expectations of interested parties. The organization needs to review who are the interested parties related to its environmental compliance obligations: customers, neighbors, and community. The organization can then consider establishing voluntary compliance requirements that could impact the environment. Once it "volunteers," the organization needs to live up to that commitment. Examples include life-cycle considerations related to product disposal, voluntary labeling on products, and environmental commitments to reduce unregulated materials (e.g., Styrofoam) or to maintain adjacent community land or waterways.

The context of the organization and expectations of interested parties will be more applicable to large, multisite organizations with transportation-related air releases—environmental impact or consumer products implication. Many organizations under ISO 14001:2004 considered greenhouse gases and climate change as part of their commitment to prevention of pollution. External factors—such as legacy issues (e.g., prior site ownership soil or water contamination responsibility)—should always be part of the organization's responsibility to ensure environmental controls are in place.

At a minimum, the organization needs to present documented evidence that the context of the organization and expectations of interested parties were considered when establishing its EMS. Some considerations for various types of organizations are listed in Table 4.1.

Table 4.1   Examples of the context and interested parties for various types of organizations.

Type of organization	Context	Interested parties
Consumer products	Climate change—transportation CO_2	Customers: life cycle—product disposal, voluntary labeling, eco-friendly materials
Multinational—contract manufacturer	Climate change—transportation CO_2	Community: landfill reduction
Chemical/materials processing	Sustainable resources	Community outreach—emergency planning
Discrete manufacturer: machining, plastic forming, casting, stamping, etc.	Sustainable resources	Community: landfill reduction

5

Clause 5: Leadership

5.1 Leadership and commitment

Top management needs to demonstrate leadership and commitment with respect to the environmental management system to ensure:

- The environmental policy and environmental objectives are established and compatible with the strategic direction and context of the organization
- The integration of the environmental management system requirements into the organization's business processes
- The resources needed for the environmental management system are available

Clause 5 is the overarching statement of what is required of top management to support the organization's EMS. The organization's management notes and performance records should indicate how effectively top management is leading the EMS by way of providing resources, strategic direction, communications, and results.

What's changed from ISO 14001:2004? **ISO 14001:2015** emphasizes management's need to support the EMS and integrate it into the organization's business planning and strategy. Later chapters offer examples where the organization's management notes can provide evidence of the integration of the EMS with the business planning and strategy.

Demonstrating leadership and commitment with respect to the EMS is an example of an ISO requirement subject to interpretation. As a third-party lead auditor providing hundreds of ISO 14001 audits, I have experienced only one organization—a large manufacturing site—where I had to issue a nonconformance against the organization's leadership. During the first two days of the audit, several issues were raised: environmental management meetings were not held per the organization's procedure, waste materials were not being stored according to procedure, and environmental objectives were not established for the current period. Clearly, top management was not supporting the EMS.

5.2 ENVIRONMENTAL POLICY

5.2 Environmental policy

Top management needs to establish, implement, and maintain an environmental policy that is appropriate to the purpose and context of the organization and provides a framework for setting environmental objectives.

The environmental policy should include:

- A commitment to the protection of the environment and prevention of pollution
- A commitment to fulfill its compliance obligations
- A commitment to continually improve its environmental management system

"Protecting the environment" can include sustainable resource use, climate change mitigation and adaptation, and protection of biodiversity and ecosystems.

The environmental policy needs to be communicated within the organization and made available to interested parties.

The organization needs to establish a documented environmental policy that is communicated internally and externally. The policy should demonstrate the organization's commitment to prevention of pollution, improvement of environmental performance, and conformance to regulatory compliance. The organization's environmental policy should be part of the organization's documented information endorsed by top management. Documented information should describe how the environmental policy is communicated within the organization and to persons doing work under the organization's control and interested parties, such as the public.

ISO 14001:2015 emphasizes the organization's commitment to include the protection of the environment, along with prevention of pollution. "Protecting the environment" can include sustainable resource use, climate change mitigation and adaptation, and protection of biodiversity and ecosystems (as appropriate).

An example of an environmental policy is shown in Figure 5.1. Note that the text "We will strive to reduce the company's carbon footprint" supports that the organization is promoting environmental protection specific to the context of its business and protection of the environment.

5.3 ORGANIZATIONAL ROLES, RESPONSIBILITIES AND AUTHORITIES

5.3 Organizational roles, responsibilities and authorities

Top management needs to ensure that the responsibilities and authorities for relevant roles are assigned and communicated within the organization.

Top management needs to assign the responsibility and authority for ensuring that the environmental management system conforms to the requirements of this International Standard, including the reporting on the performance of the environmental management system.

Environmental Policy

The *Company* will conduct our operations in a way that is protective of the environment. We will maintain an environmental management system that will serve as a framework to achieve the following goals:

Regulatory Compliance

We will identify, evaluate, and comply with all applicable federal, state, and local environmental laws and environmental requirements of our customers as well as industry standards as applicable.

Prevention of Pollution

We will seek, first, to cost-effectively avoid the creation of pollution and waste from our operations, and second, to manage remaining waste through safe and responsible methods. **We will strive to reduce the company's carbon footprint.**

Conservation

We will strive to diminish our consumption of natural resources, using sustainable resources where possible. We will strive to improve our environmental performance.

Company President   July 4, 2016

Figure 5.1   Sample environmental policy.

Top management should assign responsibilities and authorities to ensure the EMS is maintained. The organization's documented information should define individual responsibility and authority for maintaining the EMS. Examples include responsibility for reporting the performance of the EMS, authority for communicating with regulatory bodies and the public, releasing hazardous waste manifests, and approving reports to regulatory bodies.

What's changed from ISO 14001:2004? **ISO 14001:2015** does not use the title "Management Representative" as previous ISO 14001 standards did. The organization can continue to use "Management Representative" as the title having certain responsibilities, but the intent of ISO 14001:2015 is to emphasize top management's responsibilities as more than delegating.

6

Clause 6: Planning

6.1 Actions to address risks and opportunities

The organization needs to establish a process to determine the risks and opportunities related to its environmental aspects and compliance obligations, and give assurance that the environmental management system can achieve its intended outcomes.

6.1.2 Environmental aspects

The organization needs to determine the environmental aspects of its activities, products, and services that it can control and those that it can influence, and their associated environmental impacts, considering a life-cycle perspective. When determining the environmental aspects, the organization should consider changes, including planned or new developments and new or modified activities, products, and services.

The organization needs to determine those aspects that can have a significant environmental impact, i.e., significant environmental aspects, by using established criteria.

Note: The Annex to ISO 14001:2015 provides guidance on the use of this International Standard. Reference is made to the Annex throughout the *Handbook* to support certain interpretations of ISO 14001:2015 clauses. Readers of the *Handbook* are encouraged to review the Annex to increase their understanding of ISO 14001:2015. In the case of clause 6.1.1, Actions to address risks and opportunities, the Annex describes 6.1.1:

> The overall intent of the processes established in 6.1.1 is to ensure that the organization is able to achieve the intended outcomes of its environmental management system, to prevent or reduce undesired effects, and to achieve continual improvement. The organization can ensure this by determining its risks and opportunities that need to be addressed and planning action to address them. These risks and opportunities can be related to environmental aspects, compliance obligations, other issues or other needs and expectations of interested parties.

Clause 6, Planning, is the foundation on which an EMS is built. During the planning stage, an organization should review all of its processes and activities having an environmental impact. The first step is to identify the environmental aspects and impacts of the organization. A good approach is to have members of the various departments list all the activities in their group having an environmental impact. The information can be organized by media or materials, for example, sources of air discharges, sources of wastewater, storage/handling of chemicals, and types of waste at the site. Next, the utilities at the site (water, gas, oil, electricity) should be defined, followed by the aspects related to the plant exterior. What are the aspects related to production processes or maintenance? Once the departments submit their inputs, the organization can produce the summary aspects-impacts list. For illustration purposes for the next few chapters, a plant that produces injection-molded parts will be used (see Figure 6.1).

The next step is to define the criteria used to determine the significant environmental aspects (SEAs). As a starting point, the severity of the possible environmental impact should be considered. Second, the probability of occurring due to the frequency of the aspect or activity being performed should be considered. A third factor often used by organizations is the effectiveness or reliability of existing controls. While the existing controls for an environmental aspect are of importance when providing total risk analysis for an organization, when evaluating SEAs, I think it is best not to include the existing controls in the calculation, as the most severe environmental consequences often happen when controls fail.

As an example, the criteria for determining significance for the injection molding operation described in Figure 6.1 are listed in Table 6.1.

Multiplying the *frequency* by the *severity* ranks the aspects. Arbitrarily setting a score of 12 or greater would allow the first pass to establish the injection molding operation's SEAs. Jones Plastics could establish two SEAs: discharge of air from the solvent paint booth and disposal of waste paint. Note that, somewhat arbitrarily, I set the severity of electricity use at 2, since the overall use of electricity may be relatively low for an injection molding plant. Operations with high electricity use may set the severity at 3 to highlight the organization's contribution to climate change (carbon footprint). Each organization should carefully review the context of its operations when establishing the severity ranking criteria to reflect its actual and potential impact on the environment.

The example used was simplified to demonstrate a baseline aspects ranking process. In practice, the Jones Plastics staff would have included more detail on the aspects-impacts list related to which aspects have regulatory implications. In practice, it is usually more effective to complete the legal requirements analysis before establishing the SEAs. Additionally, it is important for the reader to understand that the frequency-severity criteria may not be useful for all types of manufacturing processes, especially those having greater environmental impact than the injection molding plant. A large chemical plant with multiple sites around the world may use a more complex process; a government agency may want more depth in the ranking process to allow benchmarking or comparison with various operations. Following is one of the more detailed aspects ranking processes I've witnessed:

Legal or regulatory status scale: 5 = regulated, 4 = regulated in future, 3 = company policy, 2 = company practice, 1 = unregulated

Severity scale: 5 = severe/catastrophic, 4 = serious, 3 = moderate, 2 = mild, 1 = harmless

Likelihood scale: 5 = very likely, 4 = likely, 3 = moderate, 2 = low, 1 = remote

Frequency scale: 5 = continuous, 4 = repeated, 3 = regular, 2 = intermittent, 1 = seldom

Jones Plastics, Worcester, Massachusetts

The plant covers 75,000 square feet and has 150 employees on three shifts. It produces plastic parts assembled to painted steel parts. A central receiving dock receives drums of paint, steel metal, and 55-gallon drums of alcohol for cleaning. The plant exterior has a large parking lot and two acres of landscaped grass lawn. Metal racks used to ship product to customers are stored outside. A common trash compactor is located near the loading dock. Resin is received in trucks and air-conveyed into three hoppers. The plant is heated with oil from two 500-gallon above-ground storage tanks. A 150-gallon tank of diesel fuel for the emergency generator is located outside.

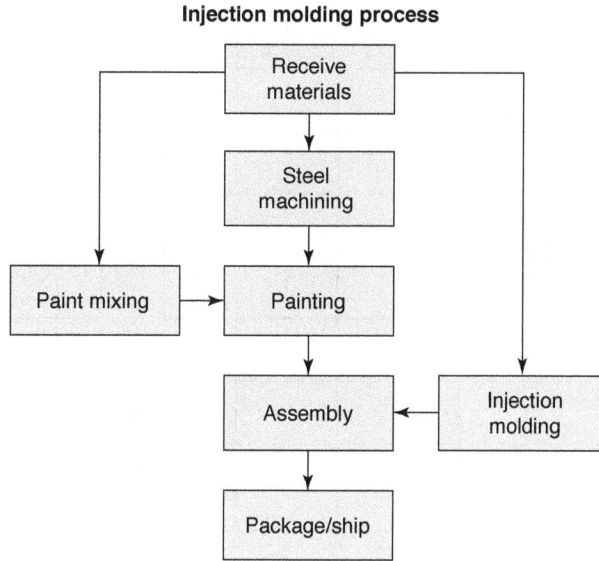

Injection molding process

Receive materials → Steel machining → Painting ← Paint mixing

Painting → Assembly ← Injection molding

Assembly → Package/ship

#	Medium	Activity/aspect	Possible environmental impact
1	Air	Solvent-based paint booth operations	Discharge of volatile organic compounds (VOCs)
2		Mix booth operation	Discharge of VOCs
3		Emergency generator—diesel	Discharge of VOCs
4		Emergency generator—diesel	Damage to groundwater or soil
5		Plant maintenance—aerosol cans	Discharge of VOCs
6		Chiller operation	Discharge of ozone-depleting substances
7		Dust filter	Discharge of particulate matter
8	Water	Roof/parking lot rainwater	Damage to groundwater or soil
9		Sanitary drains	Discharge to publicly owned treatment works
10	Waste	Disposal of mix booth waste	Land contamination
11		Disposal of bulbs/batteries	Land contamination
12		Disposal of common trash	Resource depletion/landfill use
13		Use of oil in plant and maintenance	Damage to groundwater or soil
14	Utilities	Injection molding machines	Resource depletion (electricity)
15		Chiller operation	Resource depletion (electricity)
16		Plant AC units	Resource depletion (electricity)
17	Materials	Receiving chemicals	Groundwater damage
18		Use of production resins/paints	Resource depletion (raw materials)
19		Use of packaging	Resource depletion (raw materials)
20		Office activities	Resource depletion (raw materials)
21	Plant	Heating of plant—oil boilers	Resource depletion (oil)
22		Heating of plant—oil boilers	Damage to groundwater or soil
23		Air compressors	Resource depletion (electricity)
24		Landscaping exterior property	Damage to groundwater or soil

Figure 6.1   Environmental aspects and impacts for injection molding operation example.

Table 6.1   Injection molding plant aspect rating.

#	Aspect	Frequency	Severity	Score	#	Aspect	Frequency	Severity	Score
1	Paint booths—air	5	4	20	13	Storage of oil	3	2	6
2	Mix booth—air	2	4	8	14	Molding—electricity	5	2	10
3	Emergency generator—air	2	4	8	15	Chiller—electricity	5	2	10
4	Emergency generator—fuel	2	4	8	16	AC units—electricity	5	2	10
5	Aerosol cans—air	1	4	4	17	Chemicals—water	2	4	8
6	Chillers—air	1	4	4	18	Resins/paints—resources	5	2	10
7	Dust filter—air	2	2	4	19	Packaging—resources	5	2	10
8	Parking lot water	2	4	8	20	Office activities—resources	5	2	10
9	Sanitary drains—water	5	2	10	21	Oil boilers—resources	4	2	8
10	Mix booth—waste	3	4	12	22	Oil boilers—water	2	4	8
11	Fluorescent tubes	1	4	4	23	Air compressors—electricity	5	2	10
12	Common trash	4	2	8	24	Landscaping—water	2	4	8

Controllability scale: 5 = uncontrollable, 4 = indirectly influenceable, 3 = influenceable, 2 = indirectly controllable, 1 = directly controllable

Stakeholder concerns scale: 5 = primary concern to all/most parties, 4 = primary concern to a few, 3 = secondary concern to all, 2 = secondary concern to a few, 1 = little concern

Whatever process is established, a good practice is to include in the procedure the ability for management to override the ranking results based on management judgment. An example would be an organization with a large wastewater treatment process used to remove toxic materials from its waste stream. Because this is a highly regulated process, the severity would result in a maximum score for environmental impact; however, if the process is used just a few times a year, the frequency score might yield a low total score. Management would most likely want the wastewater treatment process to be considered very significant due to the potential of a catastrophic discharge of toxic chemicals. The organization would need to prepare documented information (procedure) to define:

• The process used to identify the environmental aspects of its activities, products, and services within the defined scope of the EMS that it can

Table 6.2   Environmental aspects related to interested parties.

#	Medium	Activity/aspect	Possible environmental impact
25	Resources	Reduction of fossil based fuel sources	Reduce carbon footprint
26	Materials	Eco-friendly packaging	Landfill contamination

control and those that it can influence, taking into account planned or new developments or new or modified activities, products, and services

- The criteria for identifying those aspects that have or can have significant impacts on the environment (i.e., SEAs)

- How this information will be kept up to date

When processes are added or removed or new equipment is added, the procedure should describe how the aspects list and ranking will be maintained.

The **new requirements of ISO 14001:2015** are the actions to address risk associated with threats and opportunities. The aspects analysis and ranking process should include how the organization considers the external and internal issues that are relevant to its purpose, the needs and expectations of interested parties, and identified abnormal and potential emergency situations. The aspects-impacts chart for the injection molding plant discussed earlier could add the aspects listed in Table 6.2.

Aspect 25 involves the injection molding plant's commitment to add renewable energy sources, such as solar cells for parking lot lights, to reduce conventional electrical generation based on fossil fuels. Aspect 26 involves the plant's initiative to replace Styrofoam packaging materials with biodegradable alternatives.

The organization should consider abnormal and potential emergency situations as part of the ranking of aspects (severity). Depending on the organization's manufacturing processes, the consequences of unintended releases of chemicals, spills, explosions, and so on, are a major threat to the organization's environmental performance. Risk analysis is covered in more detail in "6.1.4 Planning Action."

6.1.3 COMPLIANCE OBLIGATIONS

> **6.1.3 Compliance obligations**
>
> The organization needs to determine and have access to the compliance obligations related to its environmental aspects and determine how these compliance obligations apply to the organization.

The next subclauses in Planning are 6.1.3, Compliance obligations, and 6.1.4, Planning action. In prior ISO 14001 revisions, these clauses were referred to as "Legal and other requirements." Returning to the EMS as a *process*, the inputs and outputs related to ISO 14001:2015 requirements are shown in Figure 6.2.

At this point in establishing the EMS, top management has led the organization to understand the context and scope of the business, endorsed the environmental

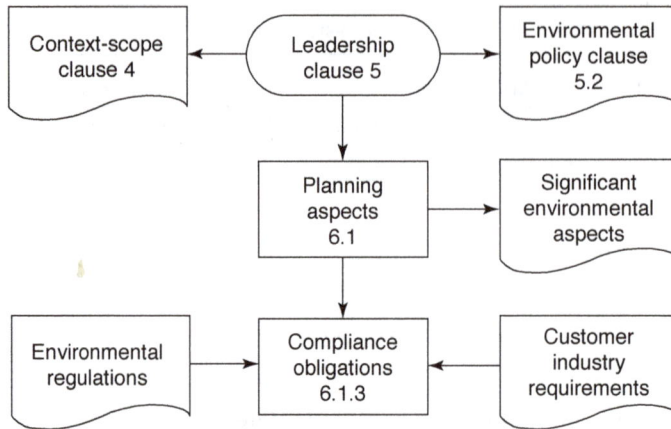

Figure 6.2   Inputs to compliance obligations.

policy, assigned responsibilities generally and responsibilities within the EMS, and supported the planning process to establish the organization's environmental aspects and impacts—including the ranking of significant aspects. The next very important process is the understanding of the organization's compliance obligations.

There are several options available to manage this process, based on the organization's environmental relevance (impact) and experience. At the high end of the relevance spectrum would be chemical processing companies, primary metals processors, and paper mills, all of which have high potential for releases to air, water, or soil; the low end would be the discrete manufacturers (machining, plastic forming, casting, stamping, etc.) with more limited environment impact.

In a high-relevance organization, the company will likely have an experienced environmental manager on staff with all the legal knowledge, permits, and records in place. If the organization is not already certified to ISO 14001, the challenge will be to convert the environmental manager's *environmental program* into an *environmental management system*. This is not a trivial point. I've provided third-party audits to organizations of all sizes, complexities, and environmental relevance. One of my biggest challenges was often the company with a strong environmental manager who openly or passively resisted working with the organization's management systems personnel. Whether it was job security protection, lack of respect for the systems people, or plain arrogance, organizations with this culture, while they maintained proper focus on compliance, often missed opportunities for improvement. The result was that during a third-party ISO 14001 audit the legal list was very difficult to audit because it was overly complex, covering every possible federal, state, and local requirement, with minimal linkage to the organization's activities. As a third-party auditor, I was responsible for encouraging the organization to conform to the ISO 14001 requirements—often by issuing nonconformances. One of the few rewards I received as a third-party auditor was when the experienced environmental manager eventually decided that the ISO 14001 environmental management system had some value!

At the low end of relevance, the organization often does not have an experienced environmental manager on staff, so it will use a consultant to manage its

Table 6.3 Regulated aspects for the injection molding plant.

#	Medium	Activity/aspect	Regulated
1	Air	Solvent-based paint booth ops.	Air discharge
2		Mix booth operation	Air discharge
3		Emergency generator—diesel	Air discharge
4		Emergency generator—diesel	Oil storage
5		Plant maintenance—aerosol cans	Air discharge
6		Chiller operation	Handling of refrigerants
7		Dust filter	Air discharges
8	Water	Roof/parking lot rainwater	Storm water
9		Sanitary drains	Wastewater
10	Waste	Disposal of mix booth waste	Hazardous waste
11		Disposal of bulbs/batteries	Universal waste
12		Disposal of common trash	Local landfill
13		Use of oil in plant and maintenance	Oil storage
14	Utilities	Injection molding machines	None
15		Chiller operation	None
16		Plant AC units	Handling of refrigerants
17	Materials	Receiving chemicals	Local fire department
18		Use of production resins/paints	None
19		Use of packaging	None
20		Office activities	None
21	Plant	Heating of plant—oil boilers	Oil storage
22		Heating of plant—oil boilers	Air discharge
23		Air compressors	None
24		Landscaping exterior property	Fertilizer

compliance obligations, under the control of the management system personnel. This combination was usually easy to audit to ISO 14001, as the compliance consultant was often experienced in ISO 14001 also.

Whether the organization is on the high or low end of environmental impact, the following process is one I've found to be successful and will conform to ISO 14001:2015. Returning to the aspects developed for the injection molding plant, Table 6.3 indicates the aspects that are regulated.

Each regulated aspect needs to be connected to a federal, state, or local requirement. It should be noted that the states can only be more restrictive than the federal government (EPA). For the example shown in Table 6.4, both EPA and Massachusetts regulations are indicated. The regulations for each media should be grouped. See Appendix E for a description of several EPA regulations.

The compliance list for the injection molding plant is listed in Table 6.5. The goal of the environmental manager should be to present this list with such clarity that the organization's top managers will be comfortable understanding the company's basic compliance requirements without the aid of the environmental

Table 6.4   Environmental regulations for the injection molding plant.

#	Aspect	Regulation
1, 2, 5	Air discharges from paint booth, mix booth emissions, aerosol cans	MassDEP, 310 CMR 7.18; EPA, 40 CFR 60 and/or 63
3, 22	Air discharges from boilers/generators	MassDEP, 310 CMR 7.02; EPA, 40 CFR 60 and/or 63
7	Air discharge—dust filter	MassDEP, 310 CMR 6.00, 7.00 and/or 8.00; EPA, 40 CFR 60
4, 13, 22	Oil storage—generator, boiler, maintenance	EPA, 40 CFR 112; MassDEP (tanks), 527 CMR 9.00
6, 15	Refrigerants—chiller, AC units	EPA, 40 CFR 82
8	Storm water—roof, parking lot	EPA National Pollutant Discharge Elimination System (NPDES), 40 CFR 122
9	Wastewater—sanitary drains	Local ordinances
10	Hazardous waste—mix booth	MassDEP, 310 CMR 30.000; EPA Resource Conservation and Recovery Act (RCRA)
11	Disposal of fluorescent tubes, batteries	MassDEP, 310 CMR 30.100; EPA 40 CFR 273
12	Disposal of common trash	Local ordinances
17	Receiving chemicals	Local ordinances
24	Landscaping	EPA—Federal Insecticide, Fungicide, and Rodenticide Act (FIFRA), 40 CFR 150–189

Table 6.5   EPA compliance list for injection molding plant.

Environmental aspect	Regulatory source	Applicable regulation	Requirement	Permit/plan
Air discharges (paint booth, mix booth, aerosol cans)	EPA, 40 CFR 60 and/or 63	To prevent, control, abate, and limit the emissions of toxic air pollutants into the ambient air	Air pollution control equipment required based on emission level of VOCs/HAPs	N/A: Maintain record of solvent use and report to EPA
Air discharges (boilers)	EPA, 40 CFR 60 and/or 63	As above		
Air discharges (dust filter)	EPA, 40 CFR 60 and/or 63	As above		
Oil storage (boiler, maintenance)	EPA, 40 CFR 112	Environmental threat posed by petroleum and non-petroleum oil spills	Spill Prevention Control & Countermeasure plan required based on amount of oil stored	SPCC dated 3/12/14
Refrigerants (chiller, AC units)	EPA, 40 CFR 82 Ozone Depleting Substances (ODS)	To define repair requirements for appliances containing refrigerants	Establishes requirements based on quantity of refrigerants used	N/A, but use licensed technician

Table 6.5   EPA compliance list for injection molding plant. (Continued)

Environmental aspect	Regulatory source	Applicable regulation	Requirement	Permit/plan
Storm water (roof, parking lot)	EPA National Pollutant Discharge Elimination System (NPDES), 40 CFR 122	To regulate discharges of pollutants into the waters of the United States	Storm water pollution prevention plan (SWPPP) required if certain materials exposed to rainwater	SWPPP dated 10/21/14
Wastewater (sanitary drains)	Local ordinances	Protect treatment plant from restricted materials	Permit	Permit 12/14/14
Hazardous waste (paint)	(RCRA) EPA, 40 CFR 260–270	Generation, storage, and disposal of hazardous waste	Requirements based on amount of hazardous waste stored	Small Quantity Generator (SQG) EPA ID# 123456
Disposal of fluorescent tubes, batteries	EPA, 40 CFR 273 Universal Waste	To regulate storage and disposal of mercury-containing devices and batteries	Requirements for labeling, protection, and disposal of universal waste	Procedure UNV 12
Disposal of common trash	EPA, 40 CFR 243.00	To regulate materials disposed in land and incinerators	Proper segregation and disposal of trash	Trash agreement 12/12/14
Receiving/ storing chemicals (55-gallon drum alcohol)	EPA, Emergency Planning & Community Right-to-Know Act (EPCRA)	To address concerns regarding the hazards posed by the storage and handling of toxic chemicals	Requires reporting based on quantity of certain chemicals used or stored	N/A
Landscaping	EPA—Federal Insecticide, Fungicide, and Rodenticide Act (FIFRA), 40 CFR 150–189	To regulate the manufacture, distribution, sale, and use of pesticides	If pesticides used, licensed applicator required	Landscaping agreement 12/12/14

Note: Readers should verify the accuracy of the regulations related to their environmental aspects from both the EPA and their particular state.

attorney. One way to achieve this is to link the aspects and regulations and explain the intent of the regulation.

In practice, this list would include additional detail. For example, "Receiving/ storing chemicals" (storage of the 55-gallon drums of alcohol) would require review with the local fire authorities, as the quantities of flammable liquid and storage protection would have restrictions. There are both federal and state regulations related to the use of toxic chemicals. The Toxic Substances Control Act (TSCA) designates PCBs, lead, asbestos, and dioxins/furans as examples of toxic materials with federal reporting requirements.

The organization needs to define and implement a process to keep the compliance list up to date. The EPA and various states have websites that can assist in that effort, or a service can be contracted. EH&S Compliance Network on LinkedIn can be a good source. There are sites with information, and possible service fees, depending on the extent of service you require—for example, Environmental Daily Advisor (http://envirodailyadvisor.blr.com/) and Aarcher Institute of Environmental Training (http://www.aarcherinstitute.com/). The *RCRA Orientation Manual 2011*, published by the EPA, is a valuable resource on the agency's solid and hazardous waste management program.

Customer, Industry Compliance Obligations

The first set of compliance obligations relates to federal, state, or local regulations; the second grouping of compliance obligations relates to restrictions placed on the organization by customers, industry, or regulatory bodies such as the European Union (EU). Familiar examples in Europe are material restrictions such as the RoHS (Restriction of Hazardous Substances) Directive and REACH (Registration, Evaluation, Authorisation and Restriction of Chemicals). ISPM 15 (International Standards for Phytosanitary Measures No. 15) is a worldwide regulation related to the composition of wood pallets and crates that addresses the need to treat wood materials used to ship products between countries. Its main purpose is to prevent the international transport and spread of disease and insects. Customers such as the automotive industry may restrict suppliers from using certain materials. Corporations with multiple plant sites often have company-wide regulations related to the use of certain materials or chemicals, for example, Styrofoam and polyvinyl chloride (PVC) packaging.

The quality department in most organizations will be the driver in understanding and conforming to customer environmental requirements. To establish a process to manage compliance with customer and industry obligations, the format used with federal regulations can be followed. As an example, the injection molding plant ships product to Mexico on wood pallets and also exports finished product to Europe for use in electronic manufacturing (see Table 6.6).

Table 6.6   Industry compliance obligations for injection molding plant.

Aspect	Regulatory source	Applicable regulation	Requirement	Comments
Shipping product to Mexico on wood pallets	International Standards for Phytosanitary Measures No. 15 (ISPM 15)	To prevent the international transport and spread of disease and insects that could negatively affect plants or ecosystems	Need to treat wood materials of a thickness greater than 6 mm that are used to ship product between countries	Use heat-treated pallets
Exporting finished product to Europe	Restriction of Hazardous Substances (RoHS) Directive	To reduce the amounts of toxic materials in electronic waste	Restricts the use of six hazardous materials in the maintenance of electronic and electrical equipment	Products certified by outside laboratory as RoHS compliant; 2/12/14

6.1.4 PLANNING ACTION (RISK ANALYSIS)

> **6.1.4 Planning action**
>
> The organization needs to plan actions to address its significant environmental aspects, compliance obligations, and risks and opportunities, and how to integrate and implement the actions into its environmental management system processes and other business processes. When planning these actions, the organization should consider its technological options and its financial, operational, and business requirements.

At this point in building the EMS, the key planning steps of establishing the environmental aspects, the SEAs, and the compliance obligations have been outlined. New to **ISO 14001:2015** is the requirement to integrate the environmental aspects and compliance obligations into the business of the organization and provide risk evaluation and planning to mitigate the risks. This clause requires an identified method to analyze risk associated with threats and opportunities related to the organization's SEAs, compliance obligations, or other issues. Results of the analysis should be used in establishing objectives and planning to mitigate the risks. The SEA analysis process should include risk assessment of aspects that present a threat to the organization's environmental results. Utilization of FMEA (failure mode effects analysis) as a quality tool could be applied. Continuing with the example of the injection molding plant, for the SEA "Discharge of air from the solvent paint booth," a risk analysis could be conducted related to a possible adverse impact on the environment during operation of the solvent paint booth (see Table 6.7).

To assess the risk related to compliance obligations, a similar process could be used (see Table 6.8).

The risks for all other regulated aspects could be assessed with severity ranking criteria established to prioritize which aspects pose the greatest threat to the organization. Appropriate actions could then be implemented to mitigate the risks. While organizations with an effective EMS certainly understand risks related to noncompliance, the new requirements of **ISO 14001:2015** in clauses 6.1.4 and 6.1.5

Table 6.7 Aspects risk assessment.

Aspect	Potential failure	Environmental impact	Present controls	New controls?
Solvent paint booth	Use of unapproved paint	Exceed VOC limits	Purchase order approval and COA (certificate of analysis)	QA certifies paint released to department
Operating solvent paint booth	Air filter not working	Discharge paint solids: neighbor complaint	Air pressure gage; measures filter quality	Add second gage and alarm
Operating solvent paint booth	Fire/explosion	Major discharge of solvent fumes	P.M. of static connections	Install in-duct gas monitor
Operating solvent paint booth	Spill of paint drum contents	Discharge to exterior of plant	Trained employees	Purchase drum containment pallets

Table 6.8 Compliance obligations risk assessment.

Regulated aspect	Potential failure	Actions to reduce exposure
Air discharges from paint/mix booths	Report not issued to MassDEP	Create compliance calendar with e-mail notifications
Air discharges from boilers and generators	Boiler malfunction	Increase P.M.
Air discharge—dust filter	Blown filter	Install gages to measure filter life
Oil storage: generator, boiler maintenance	Oil spill	Test SPCC plan more frequently Add spill kits

may have a positive effect on many organizations by formalizing the process and subjecting the "risk" evaluation process to a third-party audit.

As an extreme example, when the oil rig explosion occurred in the Gulf of Mexico in April 2010, there was much discussion concerning the role of the blowout preventer in mitigating the extent of the spill. Had the "risk" related to operation of the blowout preventer been reviewed as thoroughly as needed? Were the maintenance records up to date? There are two phases to risk analysis—the *assessment* of the risk and planning to *control or mitigate* the threat—but unfortunately the second phase is not always carried out. Often, when environmental disasters occur, an assessment is done but the follow-up control actions are not properly implemented.

Organizations with spill plans, hazardous waste contingency plans, and storm water protection requirements can use the implementation and maintenance of these plans as evidence of how they determined risks associated with threats and opportunities. By design, the Spill Prevention, Control, and Countermeasure (SPCC) rule mitigates risks related to storage of petroleum products by requiring inspections, tests, and records; personnel, training, and procedures; and drill/exercise programs. An experienced third-party ISO 14001 auditor will review the establishment, maintenance, and upgrade of an SPCC during an audit of the organization's EMS.

6.2 ENVIRONMENTAL OBJECTIVES AND PLANNING TO ACHIEVE THEM

6.2.1 Environmental objectives

The organization needs to establish environmental objectives at relevant functions and levels, taking into account its significant environmental aspects and associated compliance obligations, and considering its risks and opportunities.

The environmental objectives should be consistent with the environmental policy, measurable (if practicable), monitored, communicated, and updated as appropriate.

6.2.2 Planning actions to achieve environmental objectives

When planning how to achieve its environmental objectives, the organization needs to determine what will be done, what resources will be required, who will be responsible, and when the project will be completed.

The organization should consider how the results will be evaluated and how actions to achieve its environmental objectives can be integrated into its business processes.

Continuing with the EMS as a process (see Figure 6.3), clause 6, Planning, closes out with the requirement for the organization to establish environmental objectives and activities to achieve the objectives. In prior revisions of ISO 14001, this important step was referred to as "Objectives, targets and programs," which drove the organization to continually improve its environmental performance. In summary, ISO 14001:2015 clause 6 defines the planning required by the organization to both meet its compliance obligations and improve its environmental performance.

Often, an ISO 14001–registered organization uses only the SEAs to generate environmental objectives, overlooking the requirement to consider its technological options and financial, operational, and business requirements (as now specified in clause 6.1.4, Planning action). Graphically, the objective-setting process can be represented as in Figure 6.4.

The organization needs to establish both measurable objectives (if practicable) and a process to monitor, communicate, and update the objectives. Establishing measurable environmental objectives is fairly straightforward for most organizations. Using the "funnel" in Figure 6.4, the management staff and environmental team have several information and data sources available for the inputs of SEAs, compliance obligations, technological options, and financial, operational, and business requirements. For the injection molding plant, the analysis could be as shown in Table 6.9.

The plant's management decided to establish three objectives: reduce hazardous waste generated in the paint department, reduce plant-wide electrical use, and replace parking lot lights with solar-operated LEDs. While the discharge of air with volatile organic compounds (VOCs) from the paint booth is significant, the staff decided reductions were not a priority; however, maintaining existing controls would continue as a high priority. The finance department suggested the cost of electricity was a high percentage of operating costs—and rising. Adding the solar cells could be a subset of electrical costs and help reduce the company's carbon footprint. The sales manager liked that the solar cell project would make the company "greener" than the competition.

What situations would allow the organization to say "It is not practicable for us to set measurable goals"? The case(s) for suggesting it is not practicable (or feasible) for an organization to establish environmental objectives is usually the exception.

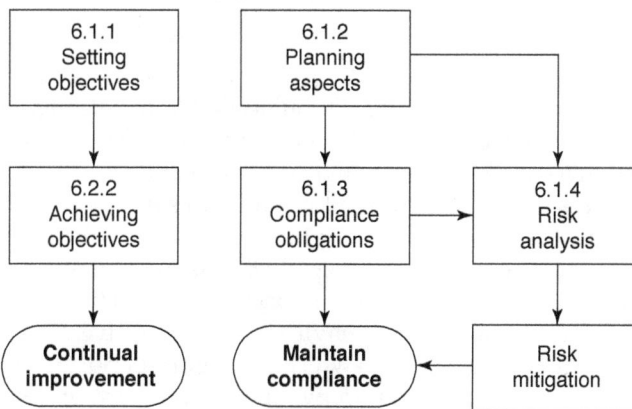

Figure 6.3   Processes to maintain compliance and improve the EMS.

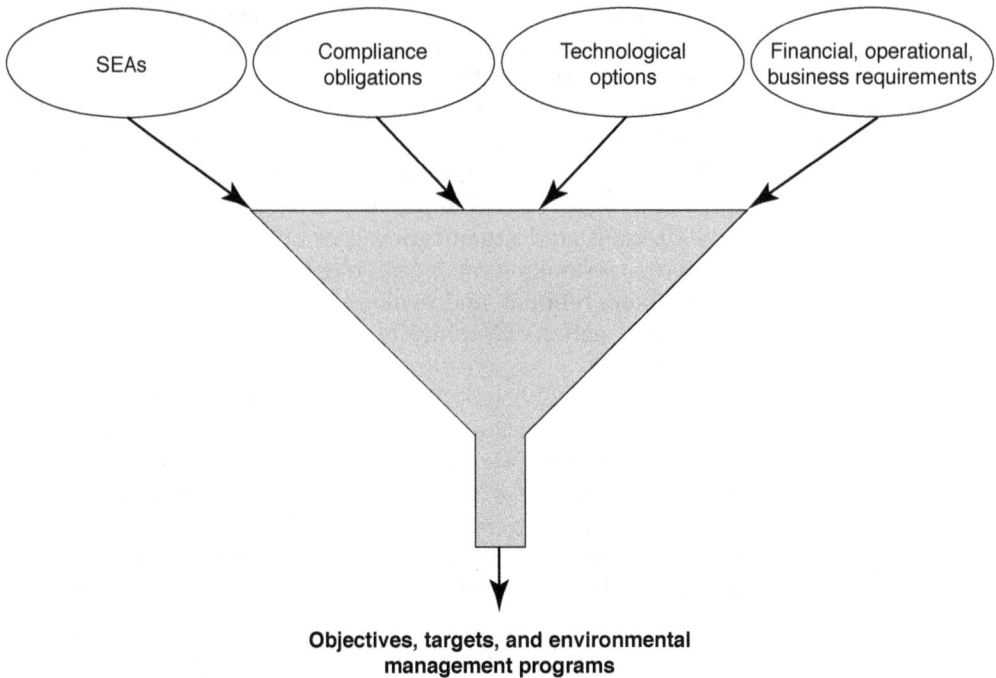

Figure 6.4   Inputs for determining the environmental objectives, targets, and programs.

Table 6.9   Potential objectives analysis.

Input	Information sources	Possible objectives
SEAs	Highest rank, risk analysis, cost, interested parties	Reduce air discharge from paint booth
Compliance obligations	Risk analysis, new regulations, customer perception	Replace toxic materials
Technological options	New developments	Add solar cells to reduce electrical costs
Financial, operational, and business requirements	Utility costs, operating costs	Reduce plant-wide use of electricity

As third-party ISO 14001 lead auditors, my colleagues and I had an unwritten guideline that suggested the organization needed to have at least one measurable objective, certainly in a manufacturing site. In some cases, a large corporation may have its headquarters (with only administrative activities) registered to ISO 14001. After a few years, HQ would run out of paper to save, or print heads to recycle, so an objective of improving response time to closing corrective actions at all the corporation's plant sites would be acceptable. On the contrary, a manufacturing site with many environmental aspects and impacts would not be treated as sympathetically. My favorite nonstarter for an environmental objective: "We will achieve registration to ISO 14001 this year." To which I would respond: "Congratulations! But show me how you are going to improve the performance of your environmental management system so I may recommend certification."

In order to monitor the objectives' performance, a target or goal needs to be established as a baseline for improvement. While establishing the environmental objectives can be a data-driven process, organizations often face challenges when establishing measurable environmental objectives; these include how to set the target when the plan is not determined and how to relate the target to changing production volumes. As a starting point, historical data from the last few years of utilities consumption, volume of wastes, use of materials, and so on, should be reviewed to form a baseline for improvement. For example, in a program such as the installation of solar cells in parking lot lights, where kilowatt-hour (kWh) consumption of the lights can be estimated, a target of X% kWh can be established. If it is not clear where the savings may come from, but there are some suggestions from experienced employees, then a range of estimates, possibly 3%–10%, can be considered.

Since electrical use in a manufacturing plant is usually related to quantity of product produced, it is recommended to *normalize* the target against a production parameter such as units shipped or sales. If production increases by 10% in the injection molding plant, electrical use may not increase directly by 10%, but there certainly will be a correlation. In some operations—for example, assembly plants with a lot of direct labor and a wide variety of finished product—the electrical use could be normalized versus the number of employees. In the case of paint waste, the correlation with production of finished injection-molded products would not be as great as the correlation with electricity used, since not all products are painted. In that case, a straight estimate can be used successfully to help drive reduction of waste.

Returning to the injection molding plant's objectives of reducing hazardous waste generated in the paint department and reducing plant-wide electrical use, the target for hazardous waste reduction, based on history, is 5%; the target for electrical use reduction is 3%.

Clause 6.2.2: Planning actions to achieve environmental objectives

Clause 6.2.2 has been described in past ISO 14001 standards as "Environmental programs." The process is basically Project Management 101: *who, what, when,* and *how*.

Continuing with the injection molding plant example, the objectives summary would be as shown in Table 6.10. By adding the program designation, the objectives and targets can now be connected to the planning requirement of clause 6.2.2.

While many options are available to indicate how the organization will plan actions to achieve environmental objectives, a template or form can be established to both plan and track the objectives with program details defined. Organizations often have project management tools and software in their business or quality systems that can be employed to manage environmental programs. Organizations of all sizes use KPIs (key process indicators) or similar to measure their business results. A best practice I have observed in many organizations is the coupling of quality objectives with environmental objectives. An example would be a reduction of waste program in a manufacturing process (quality and yield improvement) that has a collateral environmental improvement to reduce material waste sent to the landfill. This is also an example of how an organization includes improvements in environmental management in its business strategy.

Table 6.10   2015 Environmental Objectives—January 5, 2015.

Factor	Category	Objective	Target	Program
SEA	Disposal of waste paint	Reduce hazardous waste generated in the paint department	Reduce by 5% compared to FY 2014	PRG15-01
OP	Cost reduction	Reduce plant-wide electrical use	Reduce by 3% compared to FY 2014; based on kWh per mm units produced	PRG15-02
TECH	Reduction of fossil based fuel sources	Replace parking lot lights with solar-operated LEDs	100% conversion by 12/15	PRG15-03

Key: SEA = significant environmental aspect, OP = operational business, TECH = technology

---

January 10, 2015   PRG15-02

Objective: Reduce plant-wide electrical use

Target: Reduce electrical use by 3% compared to FY 2014; based on kWh per million units produced. Complete by December 31, 2015.

Project leader: Matt Sparks

#	Task	Responsibility	Schedule	Complete
1	• Develop options to reduce electrical use • Establish tracking charts	Matt	3/5/15	3/1/15
2	• Inpsect air leaks (compressors) • Repair leaks	Mike	4/1/15	
3	• Install motion sensors in offices and warehouse	Matt	6/1/15	
4	• Review options and ROI for installing solar cells and parking lot lights	Matt	6/1/15	
5	• Install solar cells for parking lot lights	Mike	12/1/15	

Figure 6.5   Example: environmental program.

For the injection molding plant, a template or form can be utilized to manage the environmental program for reducing electricity use (see Figure 6.5).

The program example in Figure 6.5 illustrates the planning, responsibilities, and resources as well as the timetable and methods to be used to achieve an environmental objective. Tracking the objectives is important in demonstrating continual improvement within the EMS.

A couple of the more interesting environmental objectives I've encountered occurred in Vermont and Brazil. A key objective for a chemical plant in Vermont was to reduce the noise generated by the trucks transporting stone from the quarry to the plant. The marble rock mine was located near a ski resort; the stone was transported some 10 miles to the manufacturing plant, where it was converted to calcium carbonate. The environmental concern was the damage the noise caused to flora and fauna in the area, but the real complaint was the lost business the owners of the ski resort inns suffered because of the noise. The program to achieve the objective involved removing the Jake Brakes from the company's fleet of trucks and replacing the mufflers (Jake Brakes change the action of the exhaust valves on a diesel engine, turning the engine into a giant and *noisy* air compressor).

The second interesting objective was at a large electronics plant in Belo Horizonte, Brazil. The company ownership was most generous to the 4000 employees at the plant by way of providing a highly subsidized lunch of wonderful barbecue and a buffet daily. For many of the employees it was the major meal of the day. Unfortunately, the "free lunch" encouraged many folks to pile more food on their plates than they would consume, overloading the plant's garbage contribution to the local landfill. The plant's objective was to reduce the average amount of food wasted per employee per month by 5 kilograms. The program to achieve the objective was centered on publicity and posters encouraging employees to be more diligent in their portion selection. It was very successful.

7

Clause 7: Support

> **7.1 Resources**
>
> The organization needs to determine and provide the resources needed for the establishment, implementation, maintenance, and continual improvement of the environmental management system.

Clauses 4–6 define the context and scope of the EMS, top management's need to demonstrate leadership and commitment in support of the EMS, and the organization's requirement to plan the activities to improve the EMS. Clause 7, Support, defines the resources needed to establish, implement, maintain, and continually improve the EMS. It includes the subclauses of competence, awareness, communications, and documented information (see Figure 7.1).

Essentially, clause 7, Support, and clause 7.1, Resources, cover requirements for how the organization will:

- Determine competence of persons doing work under its control

- Make persons doing work under the organization's control aware of the requirements of the EMS and the potential impacts associated with their work

- Plan and implement a process for internal and external communications relevant to the EMS

- Create, update, and control documented information required by the EMS and ISO 14001:2015

Clause 7.1 is the overarching requirement for the organization to support the EMS. The remaining subclauses in clause 7 provide specific requirements in competence, awareness, communications, and documented information.

> **7.2 Competence**
>
> The organization needs to determine the necessary competence of persons doing work under its control that affects its environmental performance and its ability to fulfill its compliance obligations, and ensure that these persons are competent on the basis of appropriate education, training, or experience.

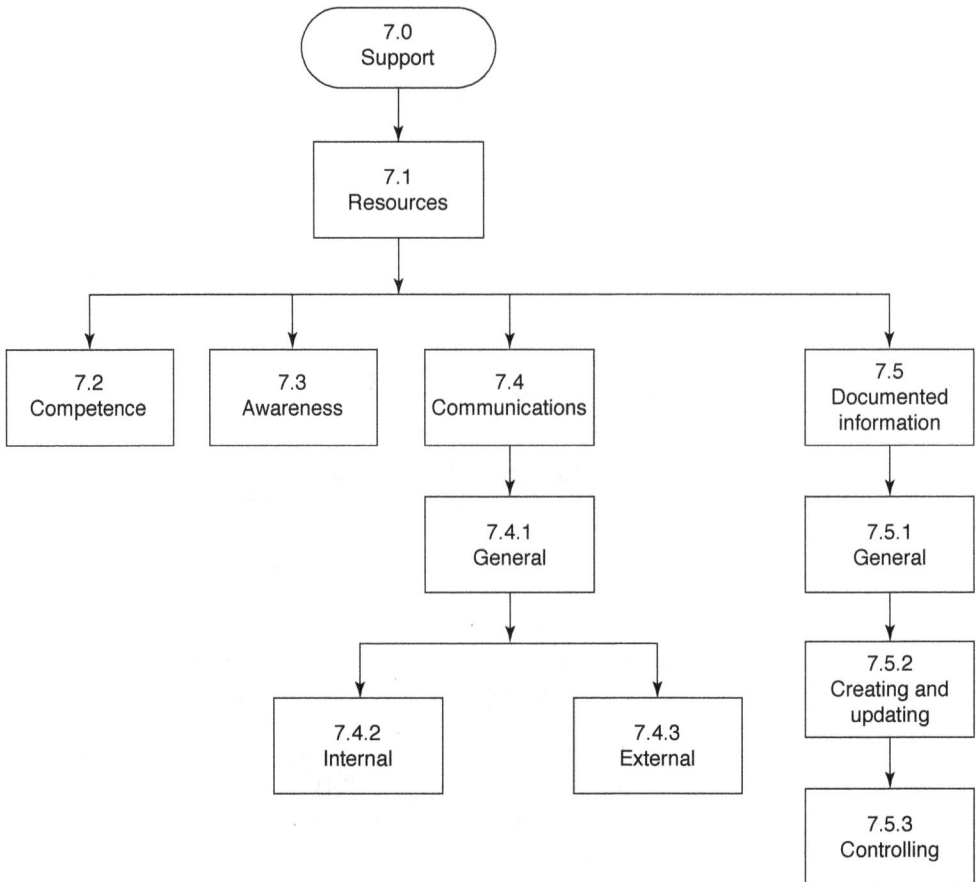

Figure 7.1 The relationship of the subclauses of clause 7, Support.

> The organization needs to determine training needs associated with its environmental aspects and its environmental management system, and where applicable, take actions to acquire the necessary competence, and evaluate the effectiveness of the actions taken.
> The organization should retain appropriate documented information as evidence of competence.

The organization needs to identify the competence, skills, and training of persons to support the EMS. The organization's environmental aspects will indicate the tasks requiring certain skill sets. Competence in the context of an industrial plant and as defined by ISO 14001:2015 is the "ability to apply knowledge and skills to achieve intended results." I prefer the Microsoft online dictionary definition: "the ability to do something well, measured against a standard, especially ability acquired through experience or training." For each environmental task, the organization needs to establish a process to qualify the assigned employees, either by witnessing the employees work, testing, or in some cases third-party licensing or certification. Where job descriptions are used, the organization should provide a record of how and why the employee matched the job's requirements. A sample

Table 7.1   Training requirements for environmental tasks.

Task	Source of training requirements	Special requirements
Wastewater treatment	Operating instructions	Local license
Handling hazardous waste	Hazardous waste contingency plan	HAZMAT/HAZWOPER
Operation of air controls	Operating instructions	EPA/state
Storm water controls	SWPPP	Certification
Spill plans	SPCC	EPA/state
Emergency team member	Emergency plans	Fire training

of environmental work assignments with competence and training requirements is shown in Table 7.1. Examples of common training requirements include:

- An organization with wastewater treatment operations would need to describe the competence requirements to operate the process; these are usually defined in the wastewater treatment operating instructions. In most cases, the state or city where the plant is located would define a license requirement, depending on the type of effluent released by the plant.

- There are training requirements for plants generating hazardous waste. Depending on the amount of hazardous waste stored by the organization, up to three forms of training may be required: Resource Conservation and Recovery Act (RCRA) training, Department of Transportation (DOT) Basic Hazmat Employee Training, and Hazardous Waste Operations and Emergency Response Standard (HAZWOPER). Initial training and refresher training need to be satisfied.

- Organizations with air releases and air controls will have, at a minimum, operating instructions to define qualifications for employees controlling or monitoring air controls. Air control devices that have high potential impact (e.g., thermal incinerator) may have EPA- or state-mandated competence requirements, particularly for employees or contractors providing maintenance and inspection of air control equipment.

- A storm water pollution prevention plan (SWPPP) requires qualified individuals to conduct inspections. (In some locations, the SWPPP inspections may have a sign-off on the form to attest to the inspector's competence to conduct the inspection.)

- An example of training employees responsible for handling oil within a Spill Prevention, Control, and Countermeasure (SPCC) could include:
  - Operation/maintenance of prevention equipment
  - Discharge procedure protocols
  - Applicable pollution control laws, rules, and regulations
  - General facility operations
  - Contents of the facility SPCC plan

- Members of emergency planning teams will require training for a wide variety of tasks such as firefighting, spill response, evacuation, weather incidents, and so forth.

In addition to the *direct* environmental impact tasks, the organization will need to provide evidence that the necessary competence of person(s) doing work under its control that affects its environmental performance is planned and documented. Examples would be receiving, handling and unloading of chemicals, segregation of waste, recycle of materials, and so forth. As an auditor, I would review all the environmental plans and operating instructions, keying in on training requirements, and then trail back to the training records maintained by either the human resources department or operating managers. During a plant audit of the various processes, I would interview employees, recording their names and tasks to later validate how each was deemed competent to perform the task.

The competence of and training for temporary workers that perform the same tasks as employees in the plant should be included in the organization's training records. Contractors hired to do special tasks—painting, electrical, plumbing work, construction, and cleaning—are covered under ISO 14001 as part of clause 8.1, Operational planning and control. This very important requirement is outlined in Chapter 8.

Chemical Right-to-Know

Under the provisions of the Hazard Communication Standard, employers are responsible for informing employees of the hazards and the identities of workplace chemicals to which they are exposed (Occupational Safety and Health Administration—OSHA 3084 1998). While hazard communications is generally considered part of the organization's safety program, many organizations combine chemical right-to-know training as part of the environmental training and awareness initiatives. Employers must establish a training and information program for employees who are exposed to hazardous chemicals in their work area at the time of initial assignment and whenever a new hazard is introduced into their work area. ISO 14001 auditors will review labeling and marking of containers to ensure that employees are aware of potential hazards and that materials are disposed of properly. The source of information is generally contained on labels on containers and material safety data sheets (MSDSs). Safety data sheets (SDSs) replaced MSDSs by mid-2016, in agreement with the United Nations' Globally Harmonized System of Classification and Labeling of Chemicals (GHS).

7.3 AWARENESS

7.3 Awareness

The organization needs to ensure that persons doing work under the organization's control are aware of the environmental policy, the significant environmental aspects and related actual or potential environmental impacts associated with their work, their contribution to the effectiveness of the environmental management system, and the implications of not conforming with the environmental management system requirements, including not fulfilling the organization's compliance obligations.

Clause 7.3, Awareness, overlaps somewhat with clause 7.2, Competence, since making employees and others aware requires some level of training; however, clause 7.3 focuses more on general understanding of the EMS, as opposed to specific tasks with impact on the organization's environmental performance, as described in clause 7.2.

ISO 14001:2015 Annex: *Guidance on the use of this International Standard* Clause A.7.3 includes this comment:

> Awareness of the environmental policy should not be taken to mean that the commitments need to be memorized or that persons doing work under the organization's control have a copy of the actual, documented environmental policy; rather, they should be aware of its existence, the purpose and their role in achieving the commitments.

This clarification is common sense in my opinion. Unfortunately, I've witnessed too many third-party auditors who expect the employees to be able to recite the environmental policy verbatim. When several employees interviewed failed to respond appropriately, nonconformances were issued. Not much value in my opinion. As long as there is evidence of management's approach to make employees aware of the environmental policy (and other facets of the EMS), clause 7.3 is satisfied. Records of attendance of ISO 14001 awareness meetings for all employees and bulletin board postings can be helpful in promoting awareness of the organization's environmental program. A best practice in various organizations I've audited is the use of television monitors in the plant that scroll information regarding the EMS; this practice gets employees' attention, particularly if the information is frequently updated.

7.4 COMMUNICATION

7.4.1 General

The organization needs to establish, implement, and maintain the processes needed for internal and external communications relevant to the environmental management system, including on what it will communicate, when to communicate, with whom to communicate, and how to communicate.

When establishing its communication processes, the organization should take into account its compliance obligations, ensuring that environmental information communicated is consistent with information generated within the environmental management system and is reliable.

The organization needs to respond to relevant communications on its environmental management system. The organization should retain documented information as evidence of its communications, as appropriate.

Clause 7.4.1 provides the overview for the organization's requirements related to communications relevant to its EMS. This clause essentially outlines the need for the organization to establish a communications procedure for internal and external environmental communications, including the requirement to retain documented information (records) of the organization's environmental communications.

> ### 7.4.2 Internal communication
>
> The organization needs to internally communicate information relevant to the environmental management system among the various levels and functions of the organization, including changes to the environmental management system, as appropriate, and ensure its communication processes enable persons doing work under the organization's control to contribute to continual improvement.
>
> ### 7.4.3 External communication
>
> The organization needs to externally communicate information relevant to the environmental management system, as established by the organization's communication processes and as required by its compliance obligations.

ISO 14001:2015 clause 7.4.1 differs from previous revisions in that it does not explicitly require the organization to make a decision on whether it will communicate its SEAs. The writers of ISO 14001:2015 emphasize the need for the organization's communications to be *reliable*. To clarify this requirement, ISO 14001:2015 Annex clause A.7.4 indicates:

Communication should:

a. be transparent, i.e. the organization is open in the way it derives what it has reported on;

b. be appropriate, so that information meets the needs of relevant interested parties', enabling them to participate;

c. be truthful and not misleading to those who rely on the information reported;

d. be factual, accurate and able to be trusted;

e. not exclude relevant information;

f. be understandable to interested parties.

Under ISO 14001:2015 clause 7.4.1 I'd be surprised (and disappointed) if a third-party auditor found reason to challenge the veracity of an organization's environmental communication. To fulfill the ISO 14001:2015 requirements for internal and external communications, the organization will need to define its plan to communicate relevant environmental issues to employees and individuals working on its behalf.

To distinguish the intent and conformance evidence among the three clauses— Competence (7.2), Awareness (7.3), and Communications (7.4)—Table 7.2 might be useful.

The competence requirements with attendant training activities are key for ensuring the success of the organization's EMS. The third-party auditors for ISO 14001 will focus on the competence level of employees performing tasks within the EMS, as errors by employees can generate the greatest risk for the organization in controlling its potential environmental unintended releases. The organization should establish a process for monitoring training and licensing requirements to ensure applicable employees maintain their competency records as required.

The awareness process for people working under the organization's control is typically audited by interviewing employees. A third-party auditor will expect the

Table 7.2   Focus and evidence for competence, awareness, and communications clauses.

#	Clause	Focus	Objective evidence
7.2	Competence	Employee/contractor ability to perform tasks within the EMS	Job descriptions; training records and licenses; compliance-related competence reports
7.3	Awareness	Employee understanding of the intent of the EMS and the environmental objectives	Awareness meeting attendance; employee interviews
7.4.2	Internal communications	Changes in the EMS; employee contributions to improvement	Postings in the plant; employee interviews
7.4.3	External communications	Ensure follow-up to external communications	Communication logs and compliance reports

employees to respond with a general understanding of the organization's environmental policy, what the policy means to them, and where the employee might find the actual environmental policy. Additionally, the auditor would expect the interviewee to have some knowledge of the organization's environmental objectives and how the employee could support the objectives.

A best practice I've witnessed for internal communication of the EMS is the use of a poster board with trend charts on the use of resources, reduction of waste, and similar performance metrics. What is not effective are poster boards that are blocked from view by equipment, charts with an unreadable small font or too much information, or worse, information that is several months out of date.

The intent of external communications is to ensure that the organization follows up on communications from neighbors, regulatory agencies, or other interested parties. An example would be a neighbor of the plant who noticed debris from the organization's trash container was blowing off the organization's property. The plant should respond to the complaint and establish corrections as appropriate. Likewise, when a regulatory agency visits the site, the event should be documented with follow-up actions defined and implemented. A useful tool to record the complaints and regulatory visits is a communications log, with personnel responsible for the responses properly trained.

As part of its external communications process, the organization should record how it will respond to requests by the public (or customers) for information related to its EMS. Will the organization provide the public with a copy of the environmental policy? Is the policy on the organization's website or other communication links? Who in the organization has the authority to report to the news media relative to an environmental incident at the site?

7.5 DOCUMENTED INFORMATION

7.5.1 General

The organization's environmental management system should include documented information required by this International Standard and documented information determined by the organization as being necessary for the effectiveness of the environmental management system.

The extent of documented information for an environmental management system should be appropriate to the size and complexity of the organization.

A nomenclature change with **ISO 14001:2015** is designating *documented information* to cover both *documents* and *records,* which were defined independently in prior ISO 14001 revisions. This is consistent with ISO 9001:2015 and is intended to allow the use of a variety of media to document the organization's plans. ISO 14001:2015 defines documented information as "information required to be controlled and maintained by an organization and the medium on which it is contained. Documented information can be in any format and media, and from any source."

7.5.2 Creating and updating

When creating and updating documented information, the organization should ensure appropriate:

- Identification and description (e.g., a title, date, author, or reference number)
- Format (e.g., language, software version, graphics) and media (e.g., paper, electronic)
- Review and approval for suitability and adequacy

7.5.3 Control of documented information

Documented information required by the environmental management system and by this International Standard should be controlled to ensure:

- It is available and suitable for use, where and when it is needed
- It is adequately protected (e.g., from loss of confidentiality, improper use, or loss of integrity)

For the control of documented information, the organization needs to address the following activities as applicable:

- Distribution, access, retrieval, and use
- Storage and preservation, including preservation of legibility
- Control of changes (e.g., version control)
- Retention and disposition

Documented information of external origin determined by the organization to be necessary for the planning and operation of the environmental management system needs to be identified, as appropriate, and controlled.

ISO 14001:2015 alleviates environmental documentation somewhat from traditional ISO 9001 quality documentation and previous ISO 14001 requirements by reducing the need for excessive document control. ISO 14001:2015 Annex A (section A.7.5) clarifies:

> Documented information originally created for purposes other than the environmental management system may be used. The documented information for the environmental management system may be integrated with other information management systems implemented by the organization. The primary focus of organizations, however, should be on the effective implementation of the environmental management system and on environmental performance, not on a complex documented information control system.

An example of where documented information differs from prior revisions of document control is the case of a supplier's operating manual, say for a thermal incinerator. In ISO 14001:2004, a third-party auditor might insist (write a nonconformance) the organization move the supplier's incinerator manual into the organization's documentation system. In ISO 14001:2015 that would not be required; the incinerator manual would be documented information of external origin, controlled as appropriate to support the EMS.

ISO 14001:2015 clause 7.5.2, Creating and updating, however, is more prescriptive than prior revisions of ISO 14001, as ISO 14001:2015 defines requirements as to how the organization should format its documented information. Procedures, work instructions, and forms need to include a title, date, author, or reference number; type of format (software version, graphics); and type of media (paper or electronic). The documented information needs to be reviewed and approved. In past revisions, the responsibilities for preparation of documents and approval were inconsistently interpreted by various organizations. The procedure describing preparation of documented information should clearly define who will prepare documents (e.g., the process owner) and who will approve documents (e.g., the environmental manager or the plant manager). A best practice would be to have a minimum of two employees of the organization sign off on each document. If a third-party consultant assists the organization in preparing ISO 14001 documentation, a member of the organization's management needs to be part of the approval process.

ISO 14001:2015 requirements for review and approval for suitability and adequacy are somewhat vague. The original issue of ISO 14001, in 1996, included the requirement "Documents are periodically reviewed, revised as necessary and approved for adequacy by authorized personnel." *Periodic* clearly indicated the need for the organization to establish a cycle time for review of environmental documents. The first revision to ISO 14001, in 2004, removed the word *periodic* to align with ISO 9001 document control terminology. In environmental documentation, I would suggest the organization establish a review process for documentation commensurate with the risks of deviation from employee instructions.

The intent of the review process is to ensure that the environmental documentation (work instructions, standard operating procedures) matches current practices. Operating personnel may improvise on how they perform a task to either improve their efficiency or save steps; this can't be allowed without management approval. The internal audit process may be a suitable method of monitoring *documentation* versus *practice*, provided audit notes include evidence that work instructions were validated. In other cases, work instructions should be reviewed by the appropriate authority at a defined frequency. My suggestion is that the work instructions related to the compliance obligations of the organization are formally reviewed annually.

ISO 14001:2015 clause 7.5.3, Control of documented information, differs from earlier ISO 14001 revisions in that *documented information* now covers what was previously referred to as *environmental records* as well as *procedures, instructions, and forms*. It is also somewhat more prescriptive than prior revisions of ISO 14001, as the organization is required to establish a process to control the distribution, access, retrieval, and use; storage and preservation, including preservation of legibility; control of changes (e.g., version control); and retention and disposition of all documented information. The majority of the requirements in practicality apply to environmental records; however, the process to control distribution, access, and version control of procedures, instructions, and forms should also be defined.

Documents of external origin determined by the organization to be necessary for the planning and operation of the EMS need to be identified, as appropriate, and controlled. Examples of external documents include legal permits, licenses, laboratory analysis, supplier equipment manuals, and customer or industry requirements or specifications. A best practice for external document control is to have a master list of all relevant external environmental documents to define location, how documents are accessed, and how they are kept current, with the revision-level control process defined.

Environmental records are a unique form of documented information. There is no requirement in ISO 14001:2015 that indicates the organization can't use the term *environmental record*. ISO 14001:2015 defines documents and records as follows: "*Document*: information created in order for the organization to operate"; "*Record*: evidence of results achieved."

Environmental records are a very important component of environmental management. In the US legal system you are innocent until proven guilty. In US environmental law, however, many environmental experts suggest that you are guilty until you can prove you are innocent. Environmental records provide that proof (either way). An experienced auditor can use the organization's listing of records to drive an effective audit. Each record will create a trail, both forward and back, of how the organization manages its environmental commitments: How was the frequency of inspections or testing established (procedure)? Why is the employee interviewed competent to do inspections (training record)? Were the discrepancies noted in the inspections resolved? What was learned in response to a chemical spill (follow-up actions)?

A sampling of environmental records present in most manufacturing organizations includes:

- Plant/area inspections

- Communications log

- Corrective actions

- Environmental incident reports

- EMS training records

- Internal audit reports

- Management review notes

- Contractor awareness

- Regulatory reports

- Compliance audit reports

- Hazardous waste manifests

- Maintenance checklists

An example of how the environmental records list might be formatted is shown in Table 7.3.

Table 7.3 Example of an environmental records list.

Record title	Form #	Location	Type	Retention	Disposition
Hazardous Waste Weekly Inspection	F-101	EHS Office	Copy	5 years	Archive
EMS Training Records	F-102	H Drive	E-File	3 years	Archive

It is helpful to include the form number as opposed to just the record title, as it assists identification. When environmental records are combined with the organization's quality records, it is best to have a clear separation of the QMS and EMS records to facilitate monitoring and auditing.

8

Clause 8: Operation

8.1 Operational planning and control

The organization needs to establish, implement, control, and maintain the processes needed to meet environmental management system requirements and establish operating criteria for the processes.

The organization needs to control planned changes and review the consequences of unintended changes, taking action to mitigate any adverse effects as necessary.

The organization needs to ensure that controls for outsourced processes are defined and implemented.

Consistent with a life-cycle perspective, the organization should establish controls, as appropriate, to ensure that its environmental requirements are addressed in the design and development process for the product or service, considering each life-cycle stage.

The organization should determine the environmental requirements for the procurement of products and services, as appropriate, and communicate its relevant environmental requirements to external providers, including contractors.

The organization should consider the need to provide information about potential significant environmental impacts associated with the transportation or delivery, use, end-of-life treatment, and final disposal of its products and services.

The organization should maintain documented information to the extent necessary to have confidence that the processes have been carried out as planned.

Clause 8 provides the requirements for operational planning and control and emergency preparedness and response within the EMS. The intent of ISO 14001:2015 clause 8 closely matches the previous revisions of ISO 14001 related to operations; however, **ISO 14001:2015** provides more specific requirements related to outsourced activities (contractors and suppliers), change control, design, and product life-cycle considerations.

Table 8.1 describes the requirements of **ISO 14001:2015** clause 8.1, with areas of focus the organization should consider as applicable to its EMS.

The design and life-cycle requirements of ISO 14001:2015 are mostly applicable to organizations manufacturing consumer products. Other organizations, such as discrete manufacturers, original equipment suppliers, and contract electronics manufacturers, are restricted by compliance obligations described in Chapter 6. If the organization has customer or industry material restrictions (e.g., RoHS [Restriction of Hazardous Substances Directive] or REACH [Registration, Evaluation, Authorisation and Restriction of Chemicals]), clause 6.1.3 requires the

Table 8.1 Focus for various requirements of clause 8.1, Operational planning and control.

ISO 14001:2015 clause 8.1 requirement	Area of focus for control
Control and maintain the processes needed to meet environmental management system requirements and establish operating criteria for the processes.	Controls need to be defined to support the organization's environmental aspects, compliance obligations, and objectives to reduce risks and support improvements.
Control planned changes and review the consequences of unintended changes, taking action to mitigate any adverse effects, as necessary.	The organization needs to ensure controls are in place when changes occur in processes, resources, and equipment.
Controls for outsourced processes are defined and implemented.	When processes are provided by contractors and/or suppliers, the environmental impacts need to be defined with appropriate controls established and implemented.
Controls, as appropriate, to ensure that its environmental requirements are addressed in the *design and development process* for the product or service, considering each life-cycle stage.	The product design process should consider the impact of materials used to avoid environmentally challenged materials during use, delivery, and disposal at end of life.
Communicate relevant environmental requirements to *external providers, including contractors.*	Contractors employed by the organization should be indoctrinated/trained as to how their activities may have impact on the organization's environmental management system.
Provide information about potential significant environmental impacts during the *delivery of the products or services and during use and end-of-life treatment of the product.*	The organization needs to provide users of its products with information related to environmental impact during use and disposal of the organization's products, as appropriate.

organization to understand these restrictions. Clause 8.1 requires the organization to demonstrate compliance with the requirements. Similarly, an organization that ships product internationally would need to demonstrate conformance to ISPM 15 (International Standards for Phytosanitary Measures No. 15), a worldwide regulation related to the composition of wood pallets and crates. A third-party auditor will sample the pallets and crates to verify that heat-treated wood or equivalent materials are being used.

Using the injection molding plant example, the SEAs might have operational controls as shown in Table 8.2.

Additionally, other aspects could have controls as listed in Table 8.3.

The intent of Table 8.3 is to demonstrate that operational controls can be a work instruction or training. When the environmental aspect is related to consumption of resources, measuring or monitoring the use of water, electricity, or gas will demonstrate control. Chapter 9 describes the requirements for monitoring.

Contractor and Supplier Operational Controls

Persons working on behalf of the organization or doing work under its control generally fall into three categories, as shown in Table 8.4.

ISO 14001:2015 has designated the requirements for these activities as outsourcing: the organization shall ensure that outsourced processes are controlled or influenced. The type and degree of control or influence to be applied to these

Table 8.2 Examples of operational controls for significant environmental aspects.

Significant environmental aspect (SEA)	Area of focus for control
Discharge of air from the solvent paint booth	Work instructions for operation, maintenance, and calibration of paint booth
Disposal of waste paint	Hazardous waste work instruction

Table 8.3 Examples of operational controls for the injection molding plant's environmental aspects.

#	Aspect	Operating controls	#	Aspect	Operating controls
1	Paint booths—air (SEA)	Work instruction	13	Storage of oil	SPCC plan
2	Mix booth—air	Training	14	Molding—electricity	Monitor
3	Emergency generator	Training	15	Chiller—electricity	Monitor
4	Emergency generator—fuel	Training	16	Plant AC units—electricity	Monitor
5	Aerosol cans—air	Work instruction	17	Chemicals—water	Work instruction
6	Chillers—air	Work instruction	18	Resins/paints—resources	Work instruction
7	Dust filter—air	Work instruction	19	Packaging—resources	Monitor
8	Parking lot rainwater	SWPPP	20	Office activities—resources	Monitor
9	Sanitary drains—water	Training	21	Oil boilers—resources	Monitor
10	Mix booth—waste (SEA)	Work instruction	22	Oil boilers—water	Monitor
11	Fluorescent tubes	Waste work instruction	23	Air compressors—electricity	Monitor
12	Common trash	Training	24	Landscaping	SWPPP

Table 8.4 Environmental activities for suppliers, temporary workers, and contractors.

Category	Type of work
Suppliers or vendors	Produce and deliver materials or services to the organization. Remove waste from the organization's site.
Temporary workers	Perform tasks to replace organization's employees.
Contractors on-site	Perform maintenance or construction or other services at organization's site.

processes shall be defined within the EMS. There is some overlap here with ISO 14001:2015 clause 7.2, Competence: the "organization shall determine the necessary competence of person(s) doing work under its control that affects its environmental performance."

When the organization hires temporary help to support its business, the competence requirements under clause 7.2 should be applied to ensure the temporary help observes the organization's environmental requirements—the same as those for equivalent company employees doing the same or similar tasks.

When the organization hires a contractor (person working under its control) to provide services such as maintenance, cleaning, painting, or construction activities, then clause 8.1 should be followed. Contracted activities are considered "outsourced processes." ISO 14001:2015 Annex A.8.1 defines an outsourced process as one that fulfills all of the following:

- The function or process is integral to the organization's functioning;

- The function or process is needed for the management system to achieve its intended outcome;

- Liability for the function or process conforming to requirements is retained by the organization; and

- The organization and the external provider have an integral relationship e.g. one where the process is perceived by interested parties as being carried out by the organization.

The organization's requirement to satisfy the ISO 14001:2015 requirement for *outsourced* processes/activities related to temporary workers, suppliers, and contractors can be described as follows:

- *Temporary workers:* Persons hired on a part-time basis to provide work to replace or supplement the organization's workforce should be trained and indoctrinated similar to employees performing the same tasks with regard to the organization's EMS (clause 7.2 applies). In large organizations, where there is often considerable use of temporary workers, the training records may reside with the temporary help agency; however, the organization needs to ensure the temporary help understands the organization's environmental requirements.

- *Suppliers or vendors:* Companies providing purchased parts or materials to the organization generally only impact their customer's environmental performance as previously described with respect to compliance obligations (e.g., RoHS for restriction on materials). The organization's supplier approval process may place requirements on how the supplier manages its environmental performance. In that case, records demonstrating conformance to a supplier's environmental commitments need to be reviewed by the organization. An updated copy of the supplier's third-party ISO 14001 certificate could be applicable.

- *Suppliers removing waste* from the organization's site have a unique and very important role in the organization's EMS. When qualifying a supplier to remove waste from its site, the organization should conduct due diligence to ensure the waste hauler has the credentials and certifications as applicable to properly execute the organization's environmental responsibilities. In the case of hazardous waste haulers, organizations will often track the supplier's process from removal at site to disposal or remediation at the supplier's site. A signed contract for all types of waste haulers is strongly recommended—often with review by the organization's legal department.

- *Contractors:* Companies or persons performing maintenance, cleaning, or construction activities at the organization's site often constitute the major outsourced activities for that organization—and likewise can have a significant impact on the organization's environmental performance. When painters, cleaners, or maintenance contractors work at the organization's site, they often bring chemicals with them as part of their work; they need to be informed of the organization's procedures for approving chemicals for use, as well as disposal restrictions. Similarly, the contractor's environmental impact on the organization's grounds, drains, and so forth, needs to be discussed and approved. To ensure the contractor has received appropriate instructions, training, and approval, a record of the orientation needs to be prepared. Organizations often create a combined safety/environmental orientation for contractors that is managed by the environmental health and safety (EHS) team.

  The environmental portion of the contractor pamphlet (or similar) should include restrictions on bringing chemicals on-site (MSDS/SDS required), disposal regulations for chemicals/materials, access to piping and tanks, and any other tasks affecting the organization's environmental performance. The organization's environmental policy can be helpful as an attachment. Additionally, the organization's SEAs can be listed, and the contractor can be asked to sign off on how their work might impact the organization's environmental performance. While the organization may choose to send the environmental orientation pamphlet to the contractor's management prior to the work, the individuals performing the work on-site should read and sign the pamphlet before starting their work.

- *Visitors* to the site: Individuals not performing contractor-type work and who are escorted by the organization's representative may receive a less detailed orientation but should be made aware of the environmental policy and applicable EHS restrictions.

During my third-party auditing experiences, a common shortfall of organizations' EMSs was contractor management. I would observe which contractors were on-site while I was auditing, and I would record the contractor's name and company for later review with the environmental coordinator. The contractor sign-in log would also be reviewed for contractors having worked on-site the previous few weeks. Too often I would find a few contractors who had not signed off as having the required EHS orientation. In some cases, the organization's entry to the site was weak, as contractors were allowed to enter via the loading dock or similar "uncontrolled" access. Review of the sign-in log would often indicate, for example, that a contractor (or visitor) had signed in on Monday but two days later there was no record that the contractor had left the site!

Note: It is strongly recommended that attention to contractor controls be considered a high-risk position in all organizations' EMSs. Many unintended environmental releases or accidents have occurred throughout the United States as a result of the lack of contractor training or inadequate communication with the contractor by the hiring organization.

8.2 EMERGENCY PREPAREDNESS AND RESPONSE

8.2 Emergency preparedness and response

The organization needs to establish, implement, and maintain the processes needed to prepare for and respond to potential emergency situations.

   The organization should:

- Prepare to respond by planning actions to prevent or mitigate adverse environmental impacts from emergency situations

- Respond to actual emergency situations

- Take action to prevent or mitigate the consequences of emergency situations, appropriate to the magnitude of the emergency and the potential environmental impact

- Periodically test the planned response actions, where practicable

- Periodically review and revise the processes and planned response actions, in particular after the occurrence of emergency situations or tests

- Provide relevant information and training related to emergency preparedness and response, as appropriate, to relevant interested parties, including persons working under its control

The organization should maintain documented information to the extent necessary to have confidence that the processes are carried out as planned.

There is some overlap for organizations related to planning for and responding to emergencies for safety and environmental incidents. It is up to the organization to determine how it will design and implement emergency preparedness and response. From the ISO 14001:2015 perspective, the requirements focus on the environmental impact once an incident has occurred. A fire in a plant may be a personnel and plant safety incident; however, the incident will most likely have environmental impact due to possible air and water contamination with sprinkler discharge interacting with materials and chemical and volatile emissions during burning. To satisfy ISO 14001 emergency preparedness and response requirements, the organization needs to establish a procedure to include the potential emergency situations and accidents that could occur at the site. Possibilities include fire, chemical release, explosion, severe weather, terrorism, and threats. The procedure should then detail how each potential incident would be considered in planning, implementation, and testing as appropriate. Most organizations, independent of their considerations in implementing an EMS, fully appreciate the need to provide emergency planning and preparedness to protect their employees and the plant. The formation of the EMS can help formalize this process, with outside audits providing monitoring. Some considerations in establishing the emergency preparedness and response plan are the following:

- How does the organization consider in its planning the environmental impact of a fire, chemical release, explosion, chemical spill, or other incident?

- Has the necessary support equipment (sprinklers, spill kits, fire extinguishers, etc.) been defined, installed, and maintained? What are the employee training requirements (records)? Is there an emergency response team? What follow-up or corrective action occurs as a result of an incident? Has the organization coordinated with local agencies (fire department, regional emergency responders)?

- Should the environmental/safety emergency planning procedures be consolidated with other emergency planning procedures? If the organization has requirements for hazardous waste management contingency planning; a Spill Prevention, Control and Countermeasure (SPCC) plan; a Storm Water Pollution Prevention Plan (SWPP); or the Emergency Planning and Community Right-to-Know ACT (EPCRA), it could be efficient to combine the procedure with the EMS emergency plan into one consolidated "master emergency" procedure. This is commonly done and referred to as an integrated contingency plan (ICP). The obvious advantage would be the avoidance of redundancy and multiple documents for maintenance, control, and training. Unfortunately, I have audited many organizations where the ICP was poorly formatted and maintained, creating difficulty in monitoring.

A large-quantity generator of hazardous waste is required to have formal written contingency plans and emergency procedures in the event of a spill or release as well as records of personnel training in the proper handling of hazardous waste through an established training program. During an audit by the EPA or state agency of an organization's hazardous waste management program, the assessor will not be impressed if he or she has to rifle through several pages of documents in the ICP to determine whether the organization complies with the requirements of a large-quantity generator.

A properly constructed ICP will provide clear separation of emergency planning components where the plan has defined regulatory commitments. Likewise, plans such as SPCC are required to have periodic review and approval by a professional engineer. The professional engineer will not want to put his or her seal on the updated SPCC if there is overlap with other somewhat related emergency planning elements. General planning for environmental/safety responses related to fire, explosion, spills, bad weather, and threats should be consolidated at the beginning of the ICP, and then tabs or dividers should be provided (if required) for SPCC, SWPPP, EPCRA, and hazardous waste contingency plans. For "change control" and revisions, each tab should have a separate cover sheet with approvals and dating. While electronic emergency planning procedures may be used, consideration needs to be given as to whether emergency instructions will be accessible if the emergency incident results in loss of power and computer networks.

Clause 8.2 sub f, Provide relevant information and training related to emergency preparedness and response, as appropriate, *to relevant interested parties*, indicates the organization should cooperate with local authorities such as fire, police, and other emergency response groups in the preplanning and monitoring of the organization's emergency planning. Depending on the potential impact of the emergency, local authorities will be better equipped to support the organization during an incident if prior review and training have occurred.

How does the organization test the emergency procedures where practicable? *Practicable* is a somewhat problematic ISO word. As defined, it means "able to be done" or "feasible." The inference in ISO 14001 is that it may not be feasible to simulate or test an actual chemical spill, fire, or explosion; however, the response activities of trained employees and all employees can be evaluated through drills. Certainly, fire alarms can be tested and are required by local agencies as well as insurance companies. Best practices I've observed with testing of emergency preparedness include an analysis of what potential emergency incidents might occur at the site (a history of incidents can assist the study) and then development of a strategy to address the risk of the emergency occurring. One year, a mock chemical spill would be simulated; other times, a full evacuation drill with execution of an emergency apparatus would be planned. In all cases, all work shifts should be considered in the plans. Most plant engineers and safety engineers reading this section will tell you the emergency "ghosts" seem to prefer working after sundown.

9

Clause 9: Performance Evaluation

9.1 MONITORING, MEASUREMENT, ANALYSIS AND EVALUATION

9.1.1 General

The organization needs to monitor, measure, analyze, and evaluate its environmental performance to determine what needs to be monitored and measured and the methods for monitoring, measurement, analysis, and evaluation, as applicable.

The criteria against which the organization will evaluate its environmental performance should be defined. The organization needs to evaluate its environmental performance and the effectiveness of the environmental management system.

The organization needs to communicate relevant environmental performance information both internally and externally.

The organization needs to ensure that calibrated or verified monitoring and measurement equipment is used and maintained, as appropriate.

Clause 9, Performance evaluation, is the "check" step in the PDCA cycle. In flowchart presentation, performance evaluation has four subsets: monitoring and measurement (objectives, operating controls, and environmental performance), evaluation of compliance, internal audit, and management review (see Figure 9.1).

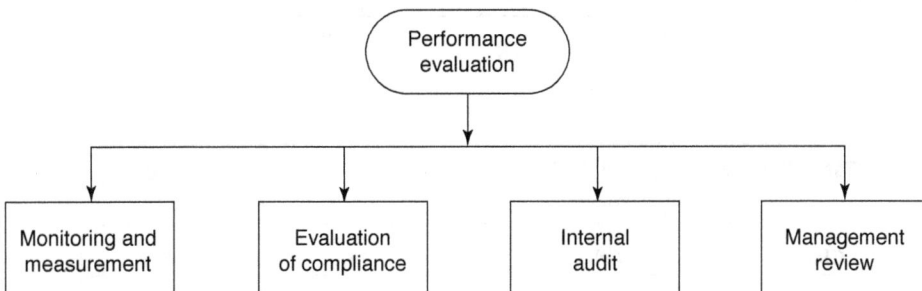

Figure 9.1 The subclauses of clause 9.0.

The organization can employ various techniques to provide evidence of its monitoring and measurement commitments. The management review notes can provide this evidence. Table 9.1 is an outline of a possible monitoring and measurement approach.

Monitoring Operating Controls

When an organization establishes an activity as an SEA, it triggers three requirements in ISO 14001:2015: did the organization consider this aspect when setting objectives, does the organization have operational controls for this aspect, and how does the organization monitor or measure this aspect? In clause 9.1, the organization needs to establish a monitoring or measurement process to ensure this aspect of high significance is controlled as planned.

Operating Controls

If one of the organization's SEAs is managing hazardous waste, there will be multiple threads in ISO 14001:2015 to monitor and audit (see Table 9.2).

When auditing clause 9.1.1, a third-party auditor will either look for evidence that *all* operational controls are monitored or measured, not just the controls related to the SEAs, or justify why the particular operational control does not require monitoring. An organization with operating controls (work instructions) established for managing receipt of chemicals will want to monitor the process using the internal audit procedure to ensure the work instruction is being followed

Table 9.1 Examples: monitoring of various EMS processes.

Category	Example	Location of evidence
Operating controls	Hazardous waste management Wastewater treatment	Hazardous waste logs, reports to agency
Objectives	Reduce hazardous waste Reduce electrical use	Charts at management review
Environmental performance	Use of electricity, gas, water, waste	Charts at management review
Calibration of instruments	pH meters, flow meters	Calibration records

Table 9.2 Monitoring of ISO 14001:2015 clauses related to managing hazardous waste.

Clause	Title	Monitoring opportunities
6.1.3	Compliance obligations	• Compliance list • EPA ID #
7.2	Competence	Training records
8.1	Operational planning and control	Work instructions
8.2	Emergency planning	Hazardous waste contingency plan
9.1	Monitoring and measurement	Hazardous waste log and manifests

as planned. An organization may have a small wastewater treatment process, which is not considered significant; however, an upset in the process may create a violation with the local authority. In this case, while the internal audit process will monitor the process for conformance to plan, tracking of wastewater quality (compared to permit) released to the agency should be included as a measurement of the effectiveness of the process.

Calibration of Environmental Monitoring Devices

Clause 9.1.1 includes the requirement "The organization needs to ensure that calibrated or verified monitoring and measurement equipment is used and maintained, as appropriate."

The rule of thumb for determining what devices require calibration in the EMS is to start with the logic: What devices support evidence that the organization is meeting its compliance obligations? An organization with a wastewater treatment obligation (and permit) to control the pH of effluent discharge to a local POTW (publicly owned treatment works) would need to ensure that the pH meter is calibrated at reasonable intervals. Experienced wastewater treatment operators, often as part of their training or certification, know how to determine the frequency of calibration. The past history of the device, extreme chemical exposure, and so on, guide the decision for the frequency of calibration. Often, pH "buffer chemicals" are used to determine pH in alkaline-neutral-acidic ranges. These solutions have a shelf life; observant auditors will look to see that the buffer solution is within its timeline to ensure accuracy of readings.

Other devices typically found in manufacturing plants include flow meters for permitted volumes of wastewater, soot measuring devices for oil-burning boilers, and heavy metal analyzers for water effluent permits. A sometimes controversial subject in environmental calibration relates to devices used to measure utilities, such as electric meters. Some third-party auditors suggest that electric meters used to track electric usage in support of energy reduction should be calibrated to ensure electric savings are accurate. I am not sure of the value of such logic, as the intent of calibrating environmental devices is to ensure that compliance obligations are met, not to manage the utility provider's responsibilities.

Thermal incinerators of all varieties demand close attention to both maintenance and calibration for the air flow parameters and temperature. One of the more unusual situations in my experiences as an auditor for an ISO 14001 registrar involved inspection of the operating controls for a thermal oxidizer at a large coating mill. In simple terms, a thermal oxidizer is employed to destroy hazardous air pollutants (HAPs) and volatile organic compounds (VOCs) from industrial air streams. To ensure the destruction efficiency meets permitting from the EPA or state authority, the air discharge temperature must stay above a specified level. On this particular audit, when reviewing the temperature recorder chart, I noted the thermal oxidizer had operated for a few hours that morning below the specified temperature by several degrees Fahrenheit. Had I uncovered the ultimate major nonconformance? Since the air discharge was in a residential neighborhood, what civic responsibility did I have in reporting the offending plant to the local authorities? After further discussion with the thermal oxidizer operator, I discovered the plant had corrected the error and informed the state permitting authority. The particular permit allowed the plant to operate the thermal oxidizer "out of condition" for three hours before being obliged to shut it down. So, the plant was not in violation, but the operator of the thermal oxidizer mentioned the plant would probably receive a visit from the state agency to review the thermal oxidizer controls and calibration record. As far as the responsibilities of auditors from outside the organization when witness to situations that could be in violation of environmental regulations, this will be discussed in the next clause, covering compliance evaluation.

9.1.2 EVALUATION OF COMPLIANCE

9.1.2 Evaluation of compliance

The organization needs to establish, implement, and maintain the processes needed to evaluate its compliance obligations by determining the frequency that compliance will be evaluated, evaluating the organization's compliance, and taking actions as needed.

The organization needs to maintain knowledge and understanding of its compliance status and retain documented information as evidence of the compliance evaluation results.

The process to evaluate conformity with compliance obligations varies greatly among organizations seeking registration to ISO 14001. Most companies with high environmental impact have compliance issues well under control prior to engaging in having their management system registered to ISO 14001. Companies with relatively low environmental impact may benefit from formalizing their compliance commitments with guidance from the ISO 14001 standard. I've formulated the following general guidelines, which come from several years' exposure to compliance auditing for organizations of all sizes and environmental intensity, for establishing an environmental compliance process that conforms to ISO 14001:2015 requirements.

Desk Top Audit

The baseline for the compliance audit should be the compliance obligations list from section 6.1.3. The first step would be a "desk top" audit of each applicable environmental aspect to verify the organization has the most current regulation in the system. Next, the related permits should be reviewed with respect to changes and the need for resubmittal to an agency. A regulation with annual or periodic reporting should be assessed. If the organization is regulated under chemical use regulations (e.g., EPCRA, TSCA), the records of annual reporting should be verified. Certain wastewater permits require quarterly reports to the local treatment plant—are they current?

Plant Tour

A plant audit (tour) should be conducted, starting with a review of the operational controls defined in Chapter 8. The standard operating procedures or similar documents will indicate where inspections or other records are required and can be verified during the plant audit. An experienced compliance auditor can easily point out discrepancies during a walk-through of the plant—inside and out. It continues to amaze me how many times I walked through a large manufacturing plant and quickly noted environmental discrepancies such as unlabeled chemical containers, spent fluorescent tubes improperly stored, hazardous waste containers with missing or inaccurate labels, and open dumpsters. During the plant tour, discrepancies should be noted and recorded in the organization's corrective action process (discussed in further detail in Chapter 10). Agreement should be established between the client and the organization as to how the compliance auditor will report issues that could result in nonconformity with its compliance obligations. Issues related to reporting environmental violations are discussed later in this chapter.

Frequency of Compliance Audits

ISO 14001:2015 requires the organization to establish the frequency with which compliance audits will be conducted. For a company seeking registration to ISO 14001 for the first time, a baseline compliance audit is required before the registrar can recommend certification. An existing compliance report can be used as long as it was done in the last year or so and provides a thorough review of the organization's compliance obligations. Each organization needs to understand the adherent risk in its operations related to environmental impact before deciding on the length of time between compliance audits. I've observed that most companies with environmental aspects related to air emissions or wastewater treatment will have a comprehensive compliance audit every third year (sometimes every fifth year), using a compliance auditor, not a member of the organization's staff. Large organizations with multiple sites often have qualified compliance auditors on the corporate staff who conduct compliance audits at each plant. More frequently, organizations will employ a third-party consultant to conduct the compliance audit. In either case, during the intervening years, it is recommended that a compliance audit be conducted each year, using an appropriate member of the organization's staff.

For smaller companies, or for larger companies with minimal environmental impact (e.g., the injection molding plant), I suggest the organization have a qualified third-party compliance auditor conduct a baseline compliance audit and then repeat every fifth year. Each intervening year, a member of the company's staff can conduct a compliance audit using a predefined checklist.

Note: When selecting a compliance auditor, it is recommended that the organization interview references supplied by the compliance auditor. Additionally, the selected compliance auditor should have direct compliance experience in the organization's state.

Attorney-Client Privilege and Notification of Violation

In the United States, compliance monitoring is one of the key components the EPA uses to ensure that the regulated community obeys environmental laws and regulations. It encompasses all regulatory agency activities performed to determine whether a facility is in compliance with applicable law. When environmental violations occur, a chain of events—sometimes quite contentious—ensues. A notification of violation (NOV) describing the incident can result in very expensive corrections, hefty fines, and even imprisonment for the offending managers or company executives. I have had several multisite auditing assignments at Fortune 500 company locations that were required to include registration to ISO 14001 as part of an imposed settlement with the EPA. In 2000, I took my lead auditor training for ISO 14001 along with several EPA lawyers who were looking to learn more about this International Standard.

A common issue involved with reviewing compliance auditing by ISO 14001 auditors is the concept of attorney-client privilege. This is a legal concept in the United States that protects certain communications between a client and their attorney and keeps those communications confidential. Organizations being audited often cite attorney-client privilege when the auditor requests to review the organization's compliance audit report. The concern from the organization is that the compliance auditor might expose an NOV that is not yet public knowledge or on the EPA's radar. ANAB (ANSI-ASQ National Accreditation Board), the organization that authorizes registrars in the United States to issue certificates to ISO 14001, has published advisories to assist auditors and registrars when faced with the attorney-client privilege challenge.

ANAB Accreditation Rule 5, issued January 1, 2014, instructs the registrars not to accept an affirmative statement in lieu of obtaining audit evidence that an organization has a satisfactory system for ensuring compliance. ANAB's instructions to the registrars have been interpreted by registrars and its auditors as follows:

> The organization needs to submit to the registrar sufficient data on the applicant's compliance with relevant legislation and regulations that are relevant and necessary to determine whether the organization's systems conform to the standard.
>   This data would include:
>
> * A documented procedure for evaluating legal compliance;
> * Evidence of its implementation, objective evidence of compliance review by management;
> * Evidence of implementation of identified corrective and preventive actions.
>
>   If any member of the audit team identifies a potential noncompliance with a legal requirement to which the organization subscribes, the potential noncompliance would be reported to the organization's management. It is expected the audited organization will use its corrective action system to investigate and correct or prevent the potential nonconformance. The potential nonconformance would not be recorded in the registrar's audit report unless the potential nonconformance is a result of a failure of the organization's environmental management system. The nonconformance would be described as a nonconformance of the requirements of the management system, and not as a compliance violation with the organization's regulatory compliance responsibilities.

Compliance auditing is an often controversial component in establishing and maintaining an EMS. Registrars and their auditors in Europe and Asia differ greatly from their counterparts in the United States on the interpretation of conformity to compliance obligations or regulatory requirements. In the United States, third-party auditors (as noted above) are trained *not to record* a potential environmental violation or NOV in their report. The reason is that the registrars and auditors do not want to get involved in litigation against their client. If the auditor's finding could indicate that the organization is not in compliance with an applicable regulation—and the violation is not yet known to the applicable enforcement agency—then the auditor could be part of the EPA's legal case against the client. The auditor will discuss the situation with the client, expecting appropriate follow-up.

Should the violation be a situation where it is both a potential regulatory violation *and* a noncompliance against the client's procedures, the auditor will issue a nonconformance designating that the client is not conforming to its requirements, and will avoid referencing legal or regulatory violation. An example I often observe involves managing hazardous waste. Generators of hazardous waste have certain requirements imposed by the EPA (and some state agencies) covering inspections, storage protocols, and training. Usually the client's procedures include controls duplicating the regulatory intention—so the EMS nonconformance will be clear. In the case where the organization does not have documented procedures duplicating the regulatory intention, will the organization avoid the auditor's nonconformance? Yes, but the organization now faces a potentially more serious nonconformance related to ISO 14001:2015:

> Clause 8.1: The organization shall maintain documented information to the extent necessary to have confidence that the processes have been carried out as planned.

Because of the lack of formalized documentation on an aspect as impactful as management of hazardous waste, a third-party auditor would now probe deeper into the organization's operational controls, possibly raising a major nonconformance in this organization's EMS.

9.2 INTERNAL AUDIT

9.2.1 General

The organization needs to conduct internal audits at planned intervals to provide information on whether the environmental management system conforms to the organization's own requirements for its environmental management system and the requirements of this International Standard, and is effectively implemented and maintained.

9.2.2 Internal audit program

The organization needs to establish, implement, and maintain an internal audit program to define the frequency, methods, planning requirements, responsibilities, and reporting of its internal audits.

When establishing the internal audit program, the organization should take into consideration the environmental importance of the processes concerned, changes affecting the organization, and the results of previous audits.

The organization needs to define the audit criteria and scope for each audit, select auditors, and conduct audits to ensure objectivity and the impartiality of the audit process and ensure that the results of the audits are reported to relevant management.

The organization should retain documented information as evidence of the implementation of the audit program and the audit results.

Many organizations seeking or holding registration to ISO 14001:2015 presently maintain certification to ISO 9001. For those organizations, the quality internal audit procedure can be extended to cover the requirements of ISO 14001. Quality auditors should be trained in the requirements of ISO 14001:2015.

In establishing an internal audit process for the EMS, the company has several requirements to address:

- What is the schedule of the audit plan?

- How is the schedule formulated?

- Have all EMS processes/clauses been audited?

- How is audit evidence obtained and recorded?

- Has an internal EMS audit team been established?

- Have the auditors been trained/qualified?

- Are audits occurring according to schedule?

- Are the follow-up actions and reporting to management timely?

An organization seeking registration to ISO 14001:2015 for the first time should provide evidence that it conducted an internal audit to all clauses shown in Figure 9.2. The internal audit results should provide information on whether the EMS conforms to the requirements of this International Standard. Additionally, the organization needs to demonstrate that it conforms to its own requirements for its EMS. To satisfy this requirement, internal audit evidence must show that the organization's practices match its interpretation of ISO 14001:2015 as well as its related documented information: procedures and instructions.

ISO 14001:2015 clauses apply to all organizations. The differences in some clauses would be the intensity of the environmental impact and related compliance obligations and operational controls. Clause 8, Operation, and clause 8.1, Operational planning and control, need to be audited specific to the organization's processes. Using the example of the injection molding plant, the internal audit plan should include the following processes: solvent-based paint booth operation, disposal of mix booth waste, receiving chemicals, disposal of bulbs/batteries, disposal of common trash, and facilities management. This organization may group the processes of operational control work instruction under management/disposal of waste, receiving chemicals, or facilities management.

The organization has a few options for how it will retain documented information as evidence of the implementation of the audit program and the audit results. Prepared check sheets or customized question lists are commonly used to drive the audit fact-finding. Organizations with a mature ISO 9001 QMS often combine the ISO 14001 internal audit with the ISO 9001 audit program, but I have found this combination to be mostly ineffective. I prefer using prepared check sheets for ISO 14001 that have been customized to match the organization's processes or environmental aspects.

A sample of the check sheet for clause 8.1, Operational planning and control, is shown in Figure 9.3.

Clause	Q1	Q2	Q3	Q4
4 Context of the organization				
5 Leadership				
6 Planning				
– 6.1.2 Environmental aspects				
– 6.1.3 Compliance obligations				
– 6.1.4 Planning action				
– 6.2 Environmental objectives and planning to achieve them				
– 6.2.1 Environmental objectives				
– 6.2.2 Environmental improvement programs				
7 Support				
– 7.1 Resources				
– 7.2 Competence				
– 7.3 Awareness				
– 7.4 Communication				
– 7.5 Documented information				
8 Operation				
– 8.1 Operational planning and control				
– 8.2 Emergency preparedness and response				
9 Performance evaluation				
– 9.1 Monitoring, measurement, analysis and evaluation				
– 9.1.1 General				
– 9.1.2 Evaluation of compliance				
– 9.2 Internal audit				
– 9.3 Management review				
10 Improvement				
– 10.1 Nonconformity and corrective action				
– 10.2 Continual improvement				

Figure 9.2 A sample audit plan.

How Should the Internal Audit Schedule Be Formulated?

My recommendation for most organizations is to conduct a full audit at least once a year. ISO 14001:2015 clause 9.2 requires that the schedule consider the environmental importance of the processes concerned, the risk associated with threats and opportunities, and the results of previous audits. The *environmental importance* and *results of previous audits* are an attempt to harmonize ISO 14001 with the requirements of the ISO 9001 quality management system standard. The QMS has a wide variety of customer-related processes: sales, design, purchasing, production, and improvement activities. In formulating an internal audit plan for the quality processes, the history of issues and importance (quality risk) can be valuable in establishing the frequency of auditing each process. If the sales process is free of

Process: Operational planning and control	Clause: 8.1
Date: _____	
Auditor: _____	Auditees: _____
	_____

C = conforming; NC = nonconforming; O = Opportunity

The organization needs to establish, implement, control and maintain the processes needed to meet environmental management system requirements, and to implement the actions identified in 6.1 (Environmental aspects) and 6.2 (Environmental objectives and planning to achieve them) by:

• establishing operating criteria for the processes;

• implementing control of the processes, in accordance with the operating criteria.

What are the environmental aspects or activities that require operational controls?	Define operating controls as applicable for each aspect:
Handling hazardous waste	
Disposal of universal waste	
Control of air emissions	
Process water treatment	
Maintenance	
Chemical handling	
Plant support equipment (air compressors, chillers, cooling towers, common trash, etc.)	

Define operating controls related to each significant environmental aspect.	
Define the operating controls in place related to outsourced processes—suppliers and contractors (as applicable).	
List the operational controls related to design and delivery of the products or services, and during use and end-of-life treatment of the product (if applicable).	
Record records related to operating controls and employees interviewed.	
Comments/NCs/opportunities:	

Figure 9.3   Sample check sheet for clause 8.1, Operational control and planning.

negative customer-related issues but the production processes have multiple customer complaints, then this organization will want to devote more time to auditing production processes.

In the EMS, I believe that the organization's greatest risks are the operational controls related to regulated environmental aspects. An organization with a thermal oxidizer, used to control air emissions, might audit the operating controls and records for the thermal oxidizer more than once a year. The same organization with a small wastewater treatment plant with an excellent history could audit that process once per year. In summary, I recommend all organizations start with an audit plan that audits all ISO 14001 clauses and environmental operational processes annually—adding more frequent audits for operational controls with high impact and risk.

What Are the Qualifications and Training Requirements for Internal Auditors?

ISO 14001:2015 clause 9.2 does not explicitly define a requirement for auditor qualifications but does specify the following: "Select auditors and conduct audits to ensure objectivity and the impartiality of the audit process." This is common practice in auditing for all management systems. An internal auditor from the operational control process should not be assigned to audit his or her own process. The individual assigned responsibility and authority to manage the EMS needs to be judicious on whether he or she gets too involved in providing internal audits for a clause he or she has responsibility to manage.

ISO 14001:2015 clause 7.2, Competence, requires the organization to "ensure that these persons are competent on the basis of appropriate education, training, or experience." An experienced third-party auditor will challenge the organization on how and why its internal auditors are qualified to conduct ISO 14001:2015 internal audits. The recommended approach for qualifying internal auditors is to have (depending on the size of the organization) a qualified EMS expert train several employees either at a public training course or on-site, if more efficient due to class size. Once a few employees are qualified to conduct ISO 14001 internal audits, the organization can have those individuals train other employees to provide internal audits. It is important that the organization have a defined plan to qualify auditors—and maintain records of how the auditors are trained. If a third-party consultant is used to conduct internal audits, the organization should keep this individual's qualifications in its files. The consultant's role should be documented in the organization's procedures.

Internal Audit Reports and Follow-Up Actions

The organization needs to maintain a report of the results of the internal audit, to include a summary report defining how the audit was conducted, issues raised (nonconformances and opportunities for improvements), and follow-up activities. Nonconformances should be entered in the organization's corrective action process. Opportunities for improvement (OFIs) should be addressed by the organization, with follow-up responses documented.

9.3 MANAGEMENT REVIEW

9.3 Management review

Top management needs to review the organization's environmental management system, at planned intervals, to ensure its continuing suitability, adequacy, and effectiveness. The management review should include the status of actions from previous management reviews and changes in:

- External and internal issues relevant to the environmental management system
- The needs and expectations of interested parties, including compliance obligations
- Its significant environmental aspects
- Risks and opportunities

The management review should include the extent to which environmental objectives have been achieved, along with information on the organization's environmental performance, including trends in:

- Nonconformities and corrective actions
- Monitoring and measurement results
- Fulfillment of its compliance obligations
- Audit results
- Adequacy of resources
- Relevant communications from interested parties, including complaints
- Opportunities for continual improvement

The outputs of the management review should include:

- Conclusions on the continuing suitability, adequacy, and effectiveness of the environmental management system
- Decisions related to continual improvement opportunities
- Decisions related to any need for changes to the environmental management system, including resources
- Actions to take if environmental objectives have not been achieved
- Opportunities to improve integration of the environmental management system with other business processes, if needed
- Any implications for the strategic direction of the organization; the organization should retain documented information as evidence of the results of management reviews

---

The organization has several options related to reporting the status of the EMS. The environmental management review meeting can be incorporated into the organization's quality management review or other business management meetings. Whatever the format, the agenda of the EMS is fairly straightforward and prescriptive—each agenda topic needs to be addressed during the frequency cycle established in the organization's planning. At a minimum, the EMS should be reviewed annually by the organization's senior staff.

Consistent with the requirement of ISO 14001:2015 clause 5.1, Leadership and commitment: ensuring the integration of the environmental management system requirements into the organization's business processes, the top management of the organization should both attend and fully participate in the environmental management meetings. Stronger emphasis on top management's involvement in the EMS is new to **ISO 14001:2015**. Many ISO 14001–certified organizations have integrated the EMS into their business model and strategy. I have audited many companies of all sizes where the environmental performance metrics are woven into the business plan: the KPIs (key process indicators) assigned to quality and business parameters also include the environmental metrics of hazardous waste reduction, material recycle, and utility use. Quality-driven waste-reduction projects include improved environmental performance.

In my observations, best-in-class organizations have established a business management system (BMS) incorporating their financial, quality, safety, and environmental systems into a cohesive operational model. A natural by-product of a successful BMS is improved employee, supplier, and community relationships. Unfortunately, I have also audited too many ISO 14001–certified organizations where top management treats the ISO 14001 certificate as "just another program" to be addressed with responsibilities delegated and managed with the least drain on resources and costs. ISO 14001:2015 is attempting to address this apparent gap by providing the registrar's third-party auditors with clear requirements related to top management's direct involvement in the organization's environmental performance and connection to the organization's business strategy.

"Suitability" of the EMS addresses whether it is appropriate for the organization's current environmental aspects and impacts—what the organization does. If the organization adds some new manufacturing or service activities, it needs to consider whether its EMS is still appropriate. "Adequacy" of the EMS refers to whether it meets the requirements of this International Standard and is implemented appropriately. "Effectiveness" refers to whether the EMS is achieving the desired results. When reviewing the EMS, management should provide a summary statement as to the suitability, adequacy, and effectiveness of the EMS, highlighting where gaps may exist and where management actions (and resources) are required to set the environmental commitments back on course.

As an agenda item, information on the organization's environmental performance should include more than reporting on the progress of the organization's performance against goals and objectives. An organization's environmental performance should include the use of resources and utilities, and waste management—recycle, environmental incidents, and compliance obligations. Members of the management team, when exposed to the trends in the organization's environmental performance, can effectively provide analysis for needs to improve the EMS using tools as described in clause 10.

10

Clause 10: Improvement

10.1 General

The organization needs to determine opportunities for improvement and implement necessary actions to achieve the intended outcomes of its environmental management system.

10.2 Nonconformity and corrective action

When nonconformity occurs, the organization needs to react to the nonconformity and, as applicable, take action to control and correct it and deal with the consequences, including mitigating adverse environmental impacts. The organization should evaluate the need for action to eliminate the causes of the nonconformity, in order that it does not recur or occur elsewhere, by reviewing the nonconformity, determining the causes of the nonconformity, and determining whether similar nonconformities exist or could potentially occur.
   The organization should:

- Implement any action needed
- Review the effectiveness of any corrective action taken
- Make changes to the environmental management system, if necessary

Corrective actions should be appropriate to the significance of the effects of the nonconformities encountered, including the environmental impacts. The organization should retain documented information as evidence of the nature of the nonconformities and any subsequent actions taken and the results of any corrective action.

10.3 Continual improvement

The organization needs to continually improve the suitability, adequacy, and effectiveness of the environmental management system to enhance environmental performance.

Clause 10, Improvement, requires the organization to establish and maintain processes to correct errors and make improvements in the EMS. In addition to ensuring the organization complies with its environmental compliance commitments, the ISO 14001:2015 standard requires the organization to analyze and improve its environmental performance. Prior versions of ISO 14001 included clause 4.5.3, Nonconformity, corrective action and preventive action: "The organization shall establish, implement and maintain a procedure(s) for dealing with actual and potential nonconformities and for taking corrective action and preventive action."

ISO 14001:2015 removes the *preventive* concept, since one of the key purposes of an EMS is to act as a preventive tool. The risk analysis clauses of ISO 14001:2015 outline a form of preventive action. The Annex to ISO 14001:2015 provides guidance for clause 10:

A.10.1 General

The organization should consider the results from analysis and evaluation of environmental performance, evaluation of compliance, internal audits and management review when taking action to improve. Examples of improvement include corrective action, continual improvement, breakthrough change, innovation and re-organization.

A.10.2 Nonconformity and corrective action

One of the key purposes of an environmental management system is to act as a preventive tool. The concept of preventive action is now captured in 4.1 (i.e. understanding the organization and its context) and 6.1 (i.e. actions to address risks and opportunities).

A.10.3 Continual improvement

The rate, extent and timescale of actions that support continual improvement are determined by the organization. Environmental performance can be enhanced by applying the environmental management system as a whole or improving one or more of its elements.

Two key processes that should support the organization's improvement activities are (1) managing the environmental programs and (2) managing corrective actions. A third-party environmental auditor will spend considerable audit time reviewing these processes. Examples of objective evidence demonstrating conformance to the subclauses in clause 10 are shown in Table 10.1.

Table 10.1 Example of objective evidence to support the organization's improvement activities.

Subclause	Source	Evidence
Environmental programs	• Performance against environmental objectives • Trend charts related to use of resources • Reduction of waste • Reduction of material sent to landfill • Replacement of toxic chemicals • Increased recycle	• Management review notes • Improvement team meeting notes • Staff business meeting notes • Employee interviews
Nonconformity and corrective actions	• Environmental incidents • Emergency drills • Compliance audits • Regulatory inspections • Neighbor complaints • Plant tours • Internal audits, third-party audits • Environmental performance	• Audit notes, findings • Tour notes • Employee suggestions • Regulator citations • Communication log

Environmental Programs

In auditing the results of an organization's improvement program, the question often arises, What if the organization has not improved its environmental performance or met its targets and objectives—is this a nonconformance? From a third-party auditor's standpoint, the *lack of results* may not constitute a nonconformance; however, the *lack of acknowledgment and actions* will often cause the auditor to issue a nonconformance, as described below.

> *Requirement:* Clause 10.1, "The organization needs to determine opportunities for improvement and implement necessary actions to achieve the intended outcomes of its environmental management system."

> *Nonconformance:* Not all environmental objectives met the established target. Reasons for failure to meet the targets or actions needed to address opportunities to improve performance are not documented.

> *Objective evidence:* Reduce hazardous waste generated in the paint department by 5% (actual 2%); reduce plant-wide electrical use by 3% (actual 1%).

Had the organization defined the reasons the objectives were not met, or had the organization issued a nonconformance during the internal audit, then a third-party auditor would not have issued a nonconformance. In the management review notes, the organization could have explained that the hazardous waste target was not met because the plant had added a new manufacturing process with major hazardous waste impact. Similarly, the electrical reduction program was also negatively impacted by the new process (more detail would be included).

Best practices used by organizations to describe the continual improvement of their EMS include detailed reporting on environmental objectives, programs and utility use, and resource management in the management review or business update meetings. Posting of the performance throughout the plant and discussions at employee meetings are also helpful to elicit support and involvement from employees.

Nonconformity and Corrective Actions

Most companies, whether or not they have a formal QMS, have a corrective action process. For over 25 years, the ISO 9001 quality management system has promulgated corrective action initiatives, which, in my opinion, have made a huge contribution to improvement in manufacturing product quality and services. In applying the corrective action process based on ISO 14001, there are several requirements—some are similar to quality, and others are unique to environmental controls.

Clause 10.2, Nonconformity and corrective action, states "Corrective actions should be appropriate to the significance of the effects of the nonconformities encountered, including the environmental impacts." There is occasion to use a formal, multifunction corrective action with cause analysis, effectiveness monitoring, and so on, and there are situations when a "find and fix" approach can be effective. During a plant environmental tour, several minor deficiencies might be observed. In these cases, rather than enter the item in the corrective action program, the issue could be "fixed" and recorded (see Table 10.2).

Table 10.2   Example: environmental issues noted during plant inspections.

August 1, 2015 plant inspection	Responsible	Repair	Follow-up	Date closed
Mix room sink leaking	John	Issue work order	Work order 98765 to shop	8/5/15
Waste drum not on containment skid	Mike	Moved onto skid	Discuss with environmental coordinator	8/2/15
Cardboard dumpster not covered	Mike	Closed	Discuss with facilities	8/2/15

The log would be reviewed by the environmental coordinator or team and kept current. Should items repeat, then a formal nonconformance/corrective action may be necessary.

Clause 10.2, Nonconformity and corrective actions, requires the organization to "evaluate the need for action to eliminate the causes of the nonconformity, in order that it does not recur or occur elsewhere, by reviewing the nonconformity, determining the causes of the nonconformity, and determining whether similar nonconformities exist or could potentially occur." In the case where the situation requires the issuance of a corrective action, the actions should not only "fix" the problem but also maximize analysis to prevent recurrence of the issue. Environmental corrective actions should include the following: correct the situation, provide analysis of cause, and provide correction for cause. The following is an example of a corrective action:

Description of nonconformity: Universal waste is not always stored according to operating procedure. Procedure Universal Waste 4/25/14 requires fluorescent tubes to be stored to protect from breakage and dated per requirements. Fluorescent tubes were found that were not dated and stored to procedure.

Corrective action: Applied proper labeling to all fluorescent bulb holders, installed lids to ensure bulbs are protected against breakage, and trained warehouse employees.

Root cause: New warehouse employee was not aware of requirements.

Correction for cause: Inspect area as waste is added to or removed from the storage area. Add posting in Universal Waste Area with picture of correct covering and dating.

To ensure that the corrections resolve the issues and avoid repeating them, review the effectiveness of any corrective action taken and make any necessary changes to the EMS. The organization should establish effectiveness measures. In this example, the organization would inspect the area with some frequency to ensure compliance with procedures. Since the root cause was the lack of training for a new employee, the organization would review the process to train employees when they are moved to a new assignment.

Most third-party auditors devote considerable audit time to reviewing the organization's corrective actions and improvement activities. When an auditor notes a discrepancy—for example, the fire alarm could not be heard in all areas of the plant during a drill—the expectation is that the organization will respond accordingly with corrections in a timely fashion.

11

Summary: Building an Environmental Management System

The ISO 14001:2015 Implementation Handbook started by describing the EMS as a "process" (see Figure 11.1):

- Top management establishes the context, scope, boundaries, and environmental policy of the EMS

- The various departments of the organization determine and rank the environmental aspects and impacts within the context/scope of the organization

- The organization determines which environmental aspects are regulated

- Operating controls are determined and implemented to ensure compliance with environmental obligations

- Objectives and programs are initiated to improve environmental performance

To supplement operational controls, employee and contractor training, emergency preparedness, and calibration of environmental devices are required. The overall EMS is supported by documentation, communication, corrective action, improvement, and management review. The internal audit and evaluation of compliance processes, along with assessment of risks, monitor the EMS to achieve the desired results of compliance with regulatory obligations and improvement of environmental performance. The development, implementation, and monitoring of all these processes will result in a successful EMS conforming to the requirements of ISO 14001:2015.

As described in the text, the International Organization for Standardization suggests organizations planning to certify to ISO 14001:2015 employ the PDCA approach for implementing the EMS. Each chapter in *The ISO 14001:2015 Implementation Handbook* links ISO 14001:2015 clauses to PDCA; these are summarized in Table 11.1.

I hope that readers of the *Handbook* have found the process approach and information and real-life examples helpful. I am available for further discussion; e-mail me at miltdentch@gmail.com. Good luck and best wishes in your quest to build or improve your EMS!

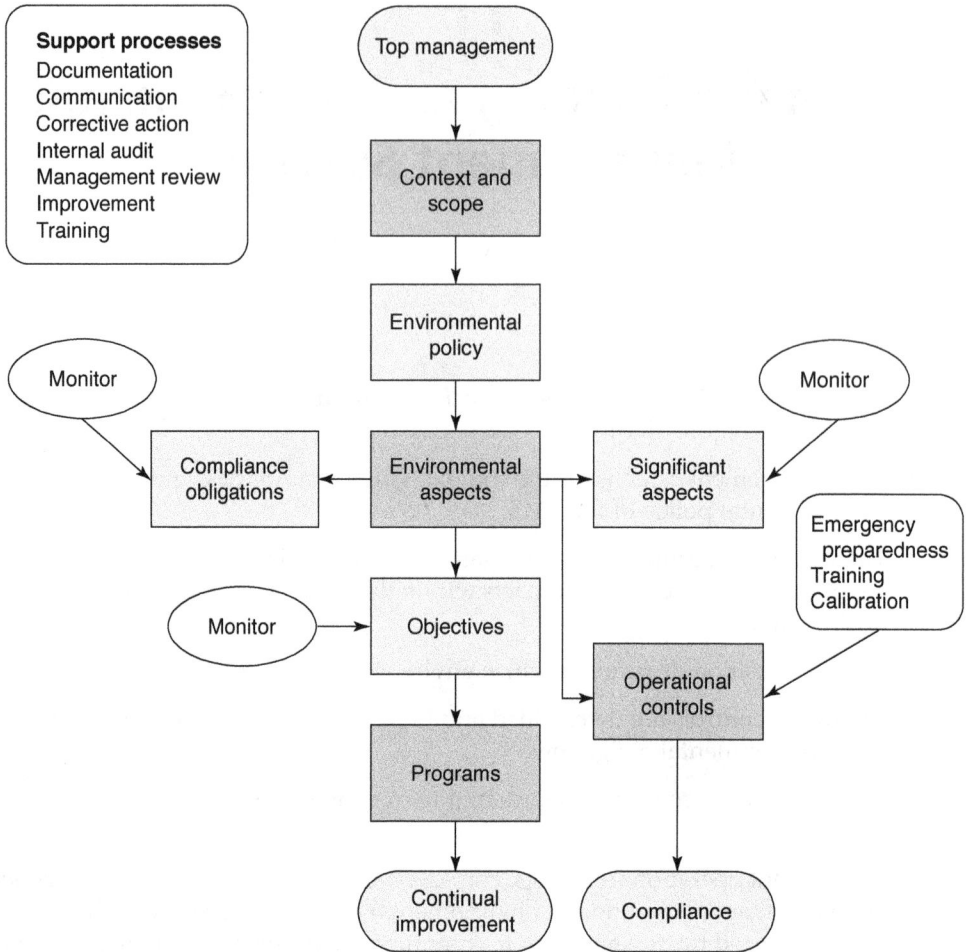

Figure 11.1 The EMS as a process.

Table 11.1 ISO 14001:2015 clauses linked to PDCA.

	Process	ISO 14001:2015	Clauses
PLAN	Review and identify the organization's environmental issues and commitments.	4 Context 5 Leadership 6 Planning	Context of the organization; Scope of the environmental management system; Leadership and commitment; Environmental policy; Environmental aspects; Compliance obligations; Objectives; Targets; Programs; Risk analysis
DO	Implement the environmental management action plans and environmental controls.	7 Support 8 Operation	Competence; Awareness; Communications; Documented information; Operational planning and control; Emergency preparedness and response
CHECK	Monitor and measure the processes and operations against the organization's objectives and report the results.	9 Performance	Monitoring, measurement, analysis and evaluation; Evaluation of compliance; Internal audit; Management review
ACT	Take actions to improve the environmental performance, making adjustments as indicated by the checks.	10 Improvement	Nonconformity and corrective action; Continual improvement

Appendix A
Correspondence: ISO 14001:2015 to ISO 14001:2004

#	ISO 14001:2015	#	ISO 14001:2004
4	**Context of the organization**	4	**Environmental management system requirements**
4.1	Understanding the organization and its context		NEW
4.2	Understanding the needs and expectations of interested parties		NEW
4.3	Determining the scope of the environmental management system	4.1	General requirements
4.4	Environmental management system	4.1	General requirements
5	**Leadership**		
5.1	Leadership and commitment	4.4.1	Resources, roles, responsibility and authority
5.2	Environmental policy	4.2	Environmental policy
5.3	Organizational roles, responsibilities and authorities	4.4.1	Resources, roles, responsibility and authority
6	**Planning**		
6.1	Actions to address risks and opportunities		
6.1.1	General		
6.1.2	Environmental aspects	4.3.1	Environmental aspects
6.1.3	Compliance obligations	4.3.2	Legal and other requirements
6.1.4	Planning action		NEW
6.2	Environmental objectives and planning to achieve them	4.3.3	Objectives, targets and programs
6.2.1	Environmental objectives	4.3.3	Objectives, targets and programs
6.2.2	Planning actions to achieve environmental objectives	4.3.3	Objectives, targets and programs
7	**Support**		
7.1	Resources	4.4.1	Resources, roles, responsibility and authority
7.2	Competence	4.4.2	Competence, training and awareness
7.3	Awareness	4.4.2	Competence, training and awareness
7.4	Communication	4.4.3	Communication
7.4.1	General	4.4.3	Communication

(continued)

81

#	ISO 14001:2015	#	ISO 14001:2004
7.4.2	Internal communication	4.4.3	Communication
7.4.3	External communication	4.4.3	Communication
7.5	Documented information	4.4.4	Documentation
7.5.1	General	4.4.4	Documentation
7.5.2	Creating and updating	4.4.4	Documentation
7.5.3	Control of documented information	4.4.5	Control of documents
		4.5.4	Control of records
8	**Operation**		
8.1	Operational planning and control	4.4.6	Operational control
8.2	Emergency preparedness and response	4.4.7	Emergency preparedness and response
9	**Performance evaluation**	**4.5**	**Checking monitoring and measurement**
9.1	Monitoring, measurement, analysis and evaluation	4.5.1	Monitoring and measuring
9.1.1	General		
9.1.2	Evaluation of compliance	4.5.2	Evaluation of compliance
9.2	Internal audit	4.5.5	Internal audit
9.2.2	Internal audit program	4.5.5	Internal audit
9.3	Management review	4.6	Management review
10	**Improvement**		
10.2	Nonconformity and corrective action	4.5.3	Nonconformity, corrective action and preventive action
10.3	Continual improvement	4.1	General requirements

Appendix B
ISO 14001 Workshop for Managers and Employees

OUTLINE

- ISO 14001 chronology
- Major elements of ISO 14001
- ISO 14001 as a process
- ISO 14001 clauses
- Role of senior management
- Common "gaps" in EMS
- Operational issues
- Best practices
- Benefits of an EMS
- Steps toward third-party registration

ISO 14001 CHRONOLOGY

- Responsible Care Program—1988
- Companies started applying BS 5750—late 1980s
- BS 7750—March 1992
- Eco-Management and Audit Scheme (EMAS)—drafted in 1992, finalized in 1995
- ISO 14001—September 1996
- ISO 14001:2004—Updated 2004
- ISO 14001:2015—Updated September 2015 (ISO 14001:2004 certificates will be withdrawn in September 2018)

ENVIRONMENTAL MANAGEMENT SYSTEM

- An EMS is that aspect of an organization's overall management structure that addresses the immediate and long-term impact of its products, services, and activities on the environment

- The goal of an EMS is to allow the organization to better understand environmental management

- It is a system, not a program

EMS CLAUSES

- Environmental policy
- Environmental aspects/impacts
- Legal and other requirements
- Significant aspects
- Objectives and targets
- Programs
- Operational control
- Emergency planning

ISO 14001: PROCESS

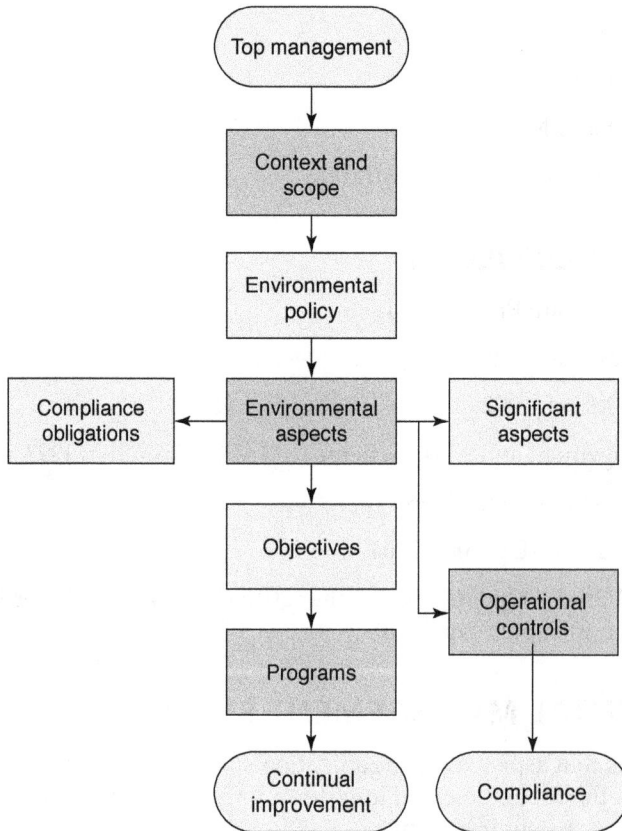

```
                    ┌─────────────────┐
                    │ Top management  │
                    └─────────────────┘
                             │
                             ▼
                    ┌─────────────────┐
                    │  Context and    │
                    │     scope       │
                    └─────────────────┘
                             │
                             ▼
                    ┌─────────────────┐
                    │  Environmental  │
                    │     policy      │
                    └─────────────────┘
                             │
                             ▼
┌──────────────┐    ┌─────────────────┐    ┌──────────────┐
│  Compliance  │◄───│  Environmental  │───►│  Significant │
│  obligations │    │     aspects     │    │   aspects    │
└──────────────┘    └─────────────────┘    └──────────────┘
                             │
                             ▼
                    ┌─────────────────┐    ┌──────────────┐
                    │   Objectives    │───►│ Operational  │
                    └─────────────────┘    │  controls    │
                             │             └──────────────┘
                             ▼                    │
                    ┌─────────────────┐           ▼
                    │    Programs     │    ┌──────────────┐
                    └─────────────────┘    │  Compliance  │
                             │             └──────────────┘
                             ▼
                    ┌─────────────────┐
                    │   Continual     │
                    │  improvement    │
                    └─────────────────┘
```

SUPPORT PROCESSES

- Documentation
- Communication
- Corrective action
- Internal audit
- Management review
- Improvement
- Training
- Risk management

ENVIRONMENTAL POLICY

- Includes a commitment to continual improvement and prevention of pollution
- Includes a commitment to comply with applicable legal requirements and with other requirements to which the organization subscribes that relate to its environmental aspects
- Is communicated to all persons working for or on behalf of the organization, and is available to the public

ASPECTS—IMPACTS

- Environmental aspect: "Element of an organization's activities or products or services that can interact with the environment" (ASQ/ANSI/ISO 14001:2015)
- Environmental impact: "Any change to the environment whether adverse or beneficial, wholly or partially resulting from an organization's environmental aspects" (ASQ/ANSI/ISO 14001:2015)

EXAMPLES OF ENVIRONMENTAL ASPECTS

- Storage of oil
- Unloading of chemicals
- Air discharge from oil-fired boiler
- Fertilizing of lawn
- Wastewater discharge from plating line
- Noise generated by trucks at loading dock
- Disposal of office waste

SIGNIFICANT ENVIRONMENTAL ASPECTS

Key step—establishes need for improvement and controls. Best practice includes involvement of members from all functions to thoroughly review aspects/impacts. Institute a ranking system to establish aspects with the highest impact on a company's environmental impact; two common factors are severity and probability of occurrence.

COMPLIANCE OBLIGATIONS

- Based on identified environmental aspects
- List of applicable regulations at federal or state level
- Develop process to maintain list as up to date
- Provide process to monitor compliance with applicable regulations
- Include requirements of other parties (corporate, industry, customer)

EXAMPLE COMPLIANCE LIST

Environmental aspect	Regulatory	Applicable regulation	Requirement	Permit/plan
Air discharges (paint booth, mix booth, aerosol cans)	MassDEP, 310 CMR 7.18 EPA, 40 CFR 60 and/or 63	To prevent, control, abate and limit the emissions of toxic air pollutants into the ambient air	If >25 tons per year of VOCs, application of air pollution control equipment required	N/A: Maintain record of solvent use and report to MassDEP
Oil storage (boiler, maintenance)	EPA, 40 CFR 112 MassDEP (tanks), 527 CMR 9.00	Potential environmental threat posed by petroleum and non-petroleum oils spills	If >1340 gallons above ground or 40,000 underground, SPCC required	SPCC dated 3/12/13

OBJECTIVES, TARGETS, AND PROGRAMS

When establishing and reviewing its objectives and targets, an organization should take into account the legal requirements and other requirements to which it subscribes, and its significant environmental aspects. It shall also consider its technological options, its financial, operational, and business requirements, and the views of interested parties.

EXAMPLE OF OBJECTIVES AND TARGETS

Objective	Target
Develop recycling program to reduce waste material to landfill	Reduce recyclable materials in waste streams to landfill by 30% by December 2014
Reduce electricity use	Reduce by 5% as percentage of units produced by December 1, 2014

PROGRAM MANAGEMENT

The organization should establish, implement, and maintain programs for achieving its objectives and targets. Programs should include:

- Designation of responsibility for achieving objectives and targets at relevant functions and levels of the organization
- The means and time frame by which they are to be achieved

OPERATIONAL CONTROL

The organization needs to identify and plan those operations that are associated with the identified significant environmental aspects consistent with its environmental policy, objectives, and targets, in order to ensure that they are carried out under specified conditions.

EXAMPLE OF OPERATING CONTROLS

Environmental aspect	Operational control
Discharge of air from the solvent paint booth	Work instructions for operation, maintenance, and calibration of paint booth
Disposal of waste paint	Hazardous waste work instruction

EMERGENCY PREPAREDNESS AND RESPONSE

- Have an emergency plan
- Identify potential situations
- Prevent or mitigate environmental impacts
- Review and revise
- Have periodic drills
- Include contingency plans for hazardous waste management as applicable

EVALUATION OF COMPLIANCE

Consistent with its commitment to compliance, the organization shall establish, implement, and maintain a procedure of periodically evaluating compliance with applicable legal and other* requirements.

*Other: corporate, customer, industry

CONFORMANCE VERSUS COMPLIANCE

- Third-party auditor assesses conformance to organization's EMS and ISO 14001
- Organization does not have to share compliance audit details with third-party auditor
- Organization is required to have process for compliance with applicable regulatory laws and regulations
- Nonconformances can be written against organization's procedures describing compliance controls

CLAUSES CONSISTENT WITH QUALITY

ISO 14001 has support requirements similar to those of ISO 9001:

- Documentation/records
- Resources/responsibilities
- Training/competence
- Communications
- Corrective/preventive actions
- Internal audit
- Management review

MANAGEMENT REVIEW

Input to management reviews should include:

- Results of internal audits and evaluations of compliance with legal requirements and with other requirements to which the organization subscribes
- Communication from external interested parties, including complaints
- The environmental performance of the organization
- The extent to which objectives and targets have been met
- Status of corrective and preventive actions
- Follow-up actions from previous management reviews

- Changing circumstances, including developments in legal and other requirements related to the organization's environmental aspects

- Recommendations for improvement

ROLE OF SENIOR MANAGEMENT

- The successful implementation of an EMS calls for a commitment from all persons working for or on behalf of the organization.

- Environmental roles and responsibilities, therefore, should not be seen as confined to the environmental management function, but can also cover other areas of an organization, such as operational management or staff functions other than environmental.

- This commitment should begin at the highest levels of management. Accordingly, top management should establish the organization's environmental policy and ensure that the EMS is implemented.

- Management should also ensure that appropriate resources, such as organizational infrastructure, are provided to ensure that the EMS is established, implemented, and maintained. Examples of organizational infrastructure include buildings, communication lines, underground tanks, and drainage.

TYPICAL "GAPS" IN EMS

- Control of and communication for "contractors"
- Objectives not measurable
- Using only SEAs to set objectives
- Too many SEAs without controls
- Controls/monitoring of SEAs not clear
- Emergency drill plan not defined
- Documentation does not match the organization's EMS

COMMON OPERATIONAL ISSUES

- Open or unmarked chemical containers
- Universal waste improperly labeled
- Inadequate control of hazardous waste
- Not linking environmental incidents to corrective action
- Lack of focus on plant exterior
- Lack of employee awareness
- Programs not detailed

BEST PRACTICES FOR EMS

- Senior management participates in EHS tours
- Contractor "sign-in" connected to SEAs
- Progress toward "zero landfill"
- Legal list is clear and connected to activities
- Compliance/activity calendar
- Trend charts for utilities/waste
- EMS programs driven by quality tools

BENEFITS OF ISO 14001

- Risk reduction
  - Environmental legal liability
  - Accidents and environmental damage
- Cost reduction
  - Disposal costs
  - Utility costs
  - Permitting fees
- Competitive edge
  - Improved corporate image
  - Investment in long-term stability
  - Improved relations with regulators
  - Counter international market pressures
  - Strategic investment now versus necessary expense later

SYSTEM VERSUS PROGRAM

Program
An environmental program:

- Can be dependent on individual knowledge
- Can be reactive, with a compliance focus only
- Can include inconsistent record keeping
- Can minimize employee involvement
- Can include a "silo" effect among managers
- Can be difficult to monitor

System (EMS)

An environmental system requires:

- Management oversight

- A commitment to improve

- Formalized record keeping

- Employee involvement

- Top management ownership and reviews

- Internal auditing

STEPS TOWARD THIRD-PARTY CERTIFICATION

- Implement EMS

  - Train employees

  - EMS audit

  - Compliance audit

  - Management review

  - Select registrar

- Stage 1—Third-party audit

  - Eliminate "gaps"

- Stage 2—Third-party audit (certification)

  - Clear nonconformances

- Receive certificate

Appendix C
Definitions from ISO 14001:2015

Audit	Systematic, independent and documented process for obtaining audit evidence and evaluating it objectively to determine the extent to which the audit criteria are fulfilled
Competence	Ability to apply knowledge and skills to achieve intended results
Compliance obligation	Requirement that an organization has to or chooses to comply with
Conformity	Fulfilment of a requirement
Continual improvement	Recurring activity to enhance performance
Corrective action	Action to eliminate the cause of a nonconformity and to prevent recurrence
Documented information	Information required to be controlled and maintained by an organization and the medium on which it is contained
Effectiveness	Extent to which planned activities are realized and planned results achieved
Environment	Surroundings in which an organization operates including air, water, land, natural resources, flora, fauna, humans and their interrelations
Environmental aspect	Element of an organization's activities or products or services that interacts or can interact with the environmental performance
Environmental condition	State or characteristic of the environment as determined at a certain point of time
Environmental impact	Change to the environment, whether adverse or beneficial, wholly or partially resulting from an organization's environmental aspects
Environmental management system	Part of the management system used to manage environmental aspects, conform to compliance obligations, and address risk associated with threats and opportunities

Environmental objective	Objective set by the organization consistent with the environmental policy
Environmental performance	Performance related to the management of environmental aspect
Environmental policy	Intentions and direction of an organization as formally expressed by its top management related to environmental performance
Indicator	Measurable representation of the condition or status of operations, management or conditions
Interested party	Person or organization that can affect, be affected by, or perceive itself to be affected by a decision or activity
Life cycle	Consecutive and interlinked stages of a product system, from raw material acquisition or generation from natural resource to end of life treatment
Management system	Set of interrelated or interacting elements of an organization to establish policies and objectives and process to achieve those objectives
Measurement	Process to determine a value
Monitoring	Determining the status of a system, a process or an activity
Nonconformity	Non-fulfilment of a requirement
Objective	Result to be achieved
Organization	Person or group of people that has its own functions with responsibilities, authorities and relationships to achieve its objectives
Outsource (verb)	Make an arrangement where an external organization performs part of an organization's function or process
Performance	Measurable result
Prevention of pollution	Use of practices, techniques, materials, products, services or energy to avoid, reduce or control (separately or in combination) the creation, emission or discharge of any type of pollutant or waste, in order to reduce adverse environmental impacts
Procedure	Specified way to carry out an activity or a process
Process	Set of interrelated or interacting activities which transforms inputs into outputs
Requirement	Need or expectation that is stated, generally implied or obligatory
Risk	Effect of uncertainty on objectives
Top management	Person or group of people who directs and controls an organization at the highest level

Appendix D
United States Environmental Protection Agency Enforcement Annual Results for Fiscal Year (FY) 2015

EPA enforcement of the nation's environmental laws is focused on large cases that drive compliance across industries and that have a high impact on protecting public health and the environment.

Our enforcement accomplishments include:

- **$7 billion** in investments by companies in actions and equipment to control pollution and clean up contaminated sites

- **$404 million** in combined federal administrative, civil judicial penalties and criminal fines

- **$4 billion** in court-ordered environmental projects resulting from criminal prosecutions

- **129 combined years** of incarceration for sentenced defendants

- **$1.975 billion** in commitments from responsible parties to clean up Superfund sites

- **$39 million** for environmental mitigation projects that provide direct benefits to local communities across the country

Enforcement Highlights

Mosaic Fertilizer, LLC, one of the world's largest fertilizer manufacturers, committed to ensuring the proper treatment, storage, and disposal of an estimated 60 billion pounds of hazardous waste at eight facilities across Florida and Louisiana, the largest amount of hazardous waste ever covered by a federal or state Resource Conservation and Recovery Act settlement.

A Clean Air Act settlement with **Hyundai-Kia** netted a $100 million fine, forfeiture of emissions credits and more than $50 million invested in compliance measures to help level the playing field for responsible companies and reduce greenhouse gas emissions fueling climate change.

Noble Energy, Inc., a leading oil and gas producer, will use advanced monitoring technologies to detect air pollution problems in real-time, and ensure proper operation and maintenance of pollution control equipment at its facilities in Colorado.

Source: United States Environmental Protection Agency, "Enforcement Annual Results for Fiscal Year (FY) 2015," http://www.epa.gov/enforcement/enforcement-annual-results-fiscal-year-fy-2015.

EPA holds criminal violators accountable that threaten the health and safety of Americans. Three subsidiaries of **Duke Energy Corporation,** the largest utility in the United States, agreed to pay a $68 million criminal fine and spend $34 million on environmental projects and land conservation to benefit rivers and wetlands in North Carolina and Virginia. As part of the plea, two Duke subsidiaries will ensure they can meet legal obligations to remediate coal ash impoundments within North Carolina, which will cost an estimated $3.4 billion.

Settlements with **Interstate Power and Light, Duke Energy Corporation** and power companies in Arizona and New Mexico are cutting coal fired power plant emissions, requiring companies to control pollution and conduct innovative projects that promote renewable energy development and energy efficiency practices.

EPA is working closely with local governments and utilities in places like **Fort Smith, Ark., Delaware County, Pa.,** and across **Puerto Rico,** to cut discharges of raw sewage and contaminated stormwater through integrated planning, green infrastructure and other innovative approaches.

Cal-Maine Foods, one of the nation's largest egg producers, is implementing a series of measures to comply with laws that control pollutants, including nutrients and bacteria, from being discharged into waterways.

XTO Energy, Inc., a subsidiary of ExxonMobil and the nation's largest holder of natural gas reserves, will restore eight sites in West Virginia damaged when streams and wetlands were filled to build roads, and implement a plan to comply with water protection laws.

Through settlements with three Nevada gold mining operations, **Newmont, Barrick and Veris,** EPA ensured that over 180 million pounds of mercury containing RCRA hazardous waste were treated, minimized, or properly disposed.

The largest bankruptcy-related cleanup settlement in American history, with **Anadarko and Kerr McGee,** will put more than $4.4 billion into toxic pollution cleanup, improving water quality and removing dangerous materials in tribal and overburdened communities.

EPA ensures federal agencies take responsibility and clean up toxic pollution. The Army addressed over 19 million cubic yards of contaminated groundwater at the **Anniston Army Depot** in Alabama, and the **U.S. Navy and Defense Logistics Agency** are required to implement at least $90 million in upgrades and improvements to prevent potential leaks at the Red Hill Bulk Storage Facility in Hawaii.

Progress on Our National Enforcement Initiatives

- Reducing Air Pollution from the Largest Sources

- Cutting Hazardous Air Pollutants

- Ensuring Energy Extraction Activities Comply with Environmental Laws

- Reducing Pollution from Mineral Processing Operations

- Keeping Raw Sewage and Contaminated Stormwater Out of Our Nation's Waters

- Preventing Animal Waste from Contaminating Surface and Ground Water

Appendix E
EPA Regulations

The following US laws and regulations are available on the EPA website (http://www.epa.gov). The notes are current as of January 2015. Readers intending to use the information provided here should verify that it is still current.

CLEAN AIR ACT

The Clean Air Act (CAA) 42 U.S.C. §7401 et seq. (1970) is the comprehensive federal law that regulates air emissions from stationary and mobile sources. Among other things, this law authorizes EPA to establish National Ambient Air Quality Standards (NAAQS) to protect public health and public welfare and to regulate emissions of hazardous air pollutants.

One of the goals of the Act was to set and achieve NAAQS in every state by 1975 in order to address the public health and welfare risks posed by certain widespread air pollutants. The setting of these pollutant standards was coupled with directing the states to develop state implementation plans (SIPs), applicable to appropriate industrial sources in the state, in order to achieve these standards. The Act was amended in 1977 and 1990 primarily to set new goals (dates) for achieving attainment of NAAQS since many areas of the country had failed to meet the deadlines.

Section 112 of the Clean Air Act addresses emissions of hazardous air pollutants. Prior to 1990, CAA established a risk-based program under which only a few standards were developed. The 1990 Clean Air Act Amendments revised Section 112 to first require issuance of technology-based standards for major sources and certain area sources. "Major sources" are defined as a stationary source or group of stationary sources that emit or have the potential to emit 10 tons per year or more of a hazardous air pollutant or 25 tons per year or more of a combination

Sources: Material is from the following web pages of the United States Environmental Protection Agency: "Summary of the Clean Air Act," http://www.epa.gov/laws-regulations/summary-clean-air-act; "Air Enforcement," http://www.epa.gov/enforcement/air-enforcement; "Water Enforcement," http://www.epa.gov/enforcement/water-enforcement; "Waste, Chemical, and Cleanup Enforcement," http://www.epa.gov/enforcement/waste-chemical-and-cleanup-enforcement; "What Is EPCRA?" http://www.epa.gov/epcra/what-epcra; "Oil Pollution Act Overview," http://archive.epa.gov/emergencies/content/lawsregs/web/html/opaover.html; "Oil Pollution Act (OPA) and Federal Facilities," http://www.epa.gov/enforcement/oil-pollution-act-opa-and-federal-facilities; "The Phase-out of Ozone-depleting Substances," http://www.epa.gov/ods-phaseout/what-phaseout-ozone-depleting-substances; and "Universal Wastes," http://www3.epa.gov/wastes/hazard/wastetypes/universal/index.htm.

of hazardous air pollutants. An "area source" is any stationary source that is not a major source.

For major sources, Section 112 requires that EPA establish emission standards that require the maximum degree of reduction in emissions of hazardous air pollutants. These emission standards are commonly referred to as "maximum achievable control technology" or "MACT" standards. Eight years after the technology-based MACT standards are issued for a source category, EPA is required to review those standards to determine whether any residual risk exists for that source category and, if necessary, revise the standards to address such risk.

Stationary Sources

Stationary sources include facilities such as factories and chemical plants, which must install pollution control equipment and meet specific emission limits under the CAA.

New Source Review (NSR) and Prevention of Significant Deterioration (PSD). These requirements require certain large industrial facilities to install state-of-the-art air pollution controls when they build new facilities or make modifications to existing facilities. Failure to install controls results in emission of pollutants that can degrade air quality and harm public health.

Reducing air pollution from the largest source of emissions is one of EPA's national enforcement initiatives. EPA is taking action to eliminate or minimize emissions from coal-fired power, acid, glass and cement plants and petroleum refineries.

- **Coal-fired power plants.** There are approximately 1,100 coal-fired electric utility units in the United States with an overall capacity of 340,000 megawatts. This sector emits approximately two-thirds of the nation's emissions inventory of sulfur dioxide (SO_2) and approximately one-third of the nitrogen oxides (NO_x). Investigations of this sector have identified a high rate of noncompliance with NSR/PSD when old plants are renovated or upgraded.

- **Plants that manufacture sulfuric and nitric acid, which are used in fertilizer, chemical and explosive production.** Acid production plants emit many thousands of tons of nitrogen oxides, sulfur dioxide, and sulfuric acid mist each year. EPA investigations have found a high rate of non-compliance with NSR/PSD in connection with plant expansions and process changes.

- **Glass manufacturing plants.** There are approximately 125 large glass plants operating in the United States. These plants emit approximately 200,000 tons per year of NO_x, SO_2 and particulate matter (PM). Investigation of this sector has shown that there have been a significant number of plant expansions but few applications for the installation of pollution controls required under NSR/PSD.

- **Cement manufacturing plants.** Cement manufacturing plants are the third largest industrial source of air pollution, emitting more than 500,000 tons per year of SO_2, NO_x and carbon monoxide. EPA determined that many cement manufacturers made changes to existing facilities without applying for and obtaining pre-construction permits. The pollution can contribute to respiratory illness and heart disease, the formation of acid rain, reduced visibility, and can be transported over long distances before falling on land or water.

- **Petroleum refineries.** Since 2000, EPA has engaged in an enforcement initiative specifically focused on addressing air emissions from petroleum refineries and has reached innovative, multi-issue, multi-facility settlement negotiations with major petroleum refining companies. These settlements have resulted in significant emission reductions of NO_x, SO_2, benzene, volatile organic compounds and PM.

Air Toxics. National Emission Standard for Hazardous Air Pollutants (NESHAP): Leaks, flares, and excess emissions from refineries, chemical plants and other industries can contain hazardous air pollutants (HAPs) that are known or suspected to cause cancer, birth defects, and seriously impact the environment. Leaking equipment is the largest source of HAP emissions from petroleum refineries and chemical manufacturing facilities. Cutting emissions of air toxics is one of EPA's National Enforcement Initiatives.

New Source Performance Standards (NSPS). Newly constructed sources or those that are modified or reconstructed must follow these standards to control excess emissions of NO_x, SO_2, and particulate matter.

Mobile Sources

Motor vehicle engines and off-road vehicles and engines must meet CAA emissions standards. These standards apply to cars, trucks, buses, recreational vehicles and engines, generators, farm and construction machines, lawn and garden equipment, marine engines and locomotives. In addition, the composition of fuels used to operate mobile sources, including gasoline, diesel, ethanol, biodiesel and blends of these fuels, are also regulated under the CAA.

New vehicles and engines must have an EPA-issued certificate of conformity before import or entry into the United States demonstrating that the engine or vehicle conforms to all applicable emissions requirements. The CAA also requires emissions labels for certified vehicles and engines.

- **Illegal imports.** Since 2008, there has been a steady flow of illegally imported uncertified motorcycles, equipment containing small gasoline-powered engines (e.g., generators, mowers, chainsaws, etc.), and recreational vehicles. Uncertified vehicles and engines can emit harmful air pollutants at 30% or more above allowable standards. EPA is working with U.S. Customs to stop illegal vehicles and engines at the ports and requiring exportation.

- **Defeat devices.** It is a violation of the CAA to manufacture, sell, or install a part for a motor vehicle that bypasses, defeats, or renders inoperative any emission control device. For example, computer software that alters diesel fuel injection timing is a defeat device. Defeat devices, which are often sold to enhance engine performance, work by disabling a vehicle's emission controls, causing air pollution. As a result of EPA enforcement, some of the largest manufacturers of defeat devices have agreed to pay penalties and stop the sale of defeat devices.

- **Tampering.** The CAA prohibits anyone from tampering with an emission control device on a motor vehicle by removing it or making it inoperable prior to or after the sale or delivery to the buyer. A vehicle's emission control system is designed to limit emissions of harmful pollutants from vehicles or engines. EPA works with manufacturers to ensure that they design their components with tamper-proofing, addresses trade groups to educate mechanics about the importance of maintaining the emission control systems, and prosecutes cases where significant or imminent harm is occurring.

Fuels. The CAA regulates fuel used in motor vehicles and non-road equipment. Clean fuels help reduce harmful emissions from a wide variety of motor vehicles, engines, and equipment.

- **Standards.** EPA regulations require that all fuel and fuel additives produced, imported and sold in the United States meet certain standards. EPA conducts targeted and random inspections to evaluate compliance with these standards, and brings enforcement actions against parties that violate these standards to reduce harmful emissions caused by fuel that does not meet the applicable standards.

- **Renewable Fuels.** Transportation fuel sold in the U.S. must contain a minimum volume of renewable fuel to reduce greenhouse gas emissions and the use of petroleum fuels. Renewable fuel producers and importers generate renewable identification number (RINs) for each gallon of renewable fuel. Refiners and importers must acquire RINs to show compliance with the standard. EPA investigates and pursues enforcement actions against anyone generating, transferring and using invalid RINs.

- **Fuel Waivers.** EPA, with the concurrence of the U.S. Department of Energy (DOE), has the authority to temporarily waive fuel or fuel additive requirements in emergency situations when the fuel supply suffers major disruptions. This helps ensure that an adequate supply of fuel is available, particularly for emergency vehicle needs. In such circumstances EPA works closely with state and other federal agencies to determine an appropriate response.

CLEAN WATER ACT

Wastewater Management. Under the CWA's National Pollutant Discharge Elimination System (NPDES) program, EPA regulates discharges of pollutants from municipal and industrial wastewater treatment plants, sewer collection systems, and stormwater discharges from industrial facilities and municipalities. The Clean Water Action Plan targets enforcement to the most important water pollution problems.

- **Municipal Wastewater and Stormwater Management.** Overflows of raw sewage and inadequately controlled stormwater discharges from municipal sewer systems can end up in waterways or cause backups into city streets or basements of homes threatening water quality, human health and the environment. EPA is targeting large municipalities to reduce pollution and volume of stormwater runoff and to reduce unlawful discharges of raw sewage that degrade water quality in communities. Reducing raw sewage overflows and stormwater discharges is one of EPA's National Enforcement Initiatives.

  - **Pretreatment.** EPA enforces requirements to ensure that industries pretreat pollutants in their wastes in order to protect local sanitary sewers and wastewater treatment plants. Industrial discharges of metals, oil and grease, and other pollutants can interfere with the operation of local sanitary sewers and wastewater treatment plants, leading to the discharge of untreated or inadequately treated pollutants into local waterways.

- **Stormwater Pollution.** This occurs when debris, chemicals, sediment or other pollutants from urban areas and construction sites get washed into storm drains and flows directly into water bodies. Uncontrolled stormwater discharges can pose significant threats to public health and the environment. The CWA requires that industrial facilities, construction sites, and municipal separate storm sewer systems (MS4s) have measures in place to prevent pollution from being discharged with stormwater into nearby waterways. Reducing discharges of contaminated stormwater into our nation's rivers, streams and lakes waterways is one of EPA's National Enforcement Initiatives.

Animal Waste and illegally discharging pollutants to water. CAFOs that are not controlling their animal wastes and illegally discharging pollutants to water bodies are a serious threat to water quality and human health. Taking action to compel these operations to properly control their wastes and comply with the law is one of EPA's National Enforcement Initiatives.

Spills—Oil and Hazardous Substances. Oil spills can harm animal and plant life, contaminating food sources and nesting habitats. Petroleum oils can form tars that persist in the environment for years. The CWA prohibits oil or hazardous substance spills in quantities that may be harmful to human health and the environment and requires actions to prevent future spills.

Wetlands—Discharges of Dredge and Fill Material. EPA ensures that dredged or fill material is not discharged into wetlands and other waters of the United States except as authorized by a permit issued by the United States Army Corps of Engineers. EPA investigates and inspects those discharging dredge and fill material into wetlands and other waters of the United States without a permit and pursues appropriate enforcement to ensure compliance.

Deepwater Horizon—BP Gulf of Mexico Oil Spill. On April 20, 2010, the oil drilling rig, *Deepwater Horizon*, operating in the Macondo Prospect in the Gulf of Mexico, exploded and sank resulting in the death of 11 workers on the *Deepwater Horizon* and the largest spill of oil in the history of marine oil drilling operations.

Clean Water Act Compliance Monitoring and Assistance

EPA works with its federal, state and tribal regulatory partners through a comprehensive Clean Water Act compliance monitoring program to protect human health and the environment by ensuring that the regulated community obeys environmental laws/regulations through on-site visits by qualified inspectors, and a review of the information EPA or a state/tribe requires to be submitted. The CWA compliance assistance program provides businesses, federal facilities, local governments and tribes with tools to help meet environmental regulatory requirements.

SAFE DRINKING WATER ACT (SDWA)

Drinking Water. EPA safeguards human health by enforcing the requirements of the SDWA to ensure that the nation's public drinking water supply and its sources (rivers, lakes, reservoirs, springs, and ground water wells) are protected.

Public Drinking Water Systems. EPA ensures that public drinking water systems comply with health-based federal standards for contaminants, which includes performing regular monitoring and reporting.

- **Underground Injection Control.** Underground injection is the technology of placing fluids underground, in porous formations of rocks, through wells or other similar conveyance systems. EPA ensures that underground injection wells do not endanger any current and future underground or surface sources of drinking water. EPA investigates and inspects those injecting fluids underground in violation of the SDWA and pursues appropriate enforcement to ensure compliance.

- **Aircraft Drinking Water Rule.** EPA is responsible for ensuring the safety of drinking water on aircraft and is working with airlines to ensure drinking water quality to include making certain that the airlines are in compliance with the recently promulgated Aircraft Drinking Water Rule.

Safe Drinking Water Act Compliance Monitoring and Assistance

EPA works with its federal, state and tribal regulatory partners through a comprehensive Safe Drinking Water Act compliance monitoring program to protect human health and the environment by ensuring that the regulated community obeys environmental laws/regulations through on-site visits by qualified inspectors, and a review of the information EPA or a state/tribe requires to be submitted. The SDWA compliance assistance program provides businesses, federal facilities, local governments and tribes with tools to help meet environmental regulatory requirements.

WASTE, CHEMICAL, AND CLEANUP ENFORCEMENT

EPA enforces a variety of environmental requirements related to pollution by waste and chemicals.

Waste Enforcement

Mining and mineral processing. EPA is taking action to protect communities and the environment from illegal or high risk hazardous waste operations at phosphoric acid and other high risk mineral processing facilities. Reducing pollution from mining and mineral processing operations is one of EPA's national initiatives.

Hazardous wastes. EPA enforces requirements under the Resource Conservation and Recovery Act regarding the safe handling, treatment, storage and disposal of hazardous wastes. EPA and the states verify RCRA compliance with these requirements through a comprehensive compliance monitoring program which includes inspecting facilities, reviewing records and taking enforcement action where necessary. The RCRA compliance assistance program provides businesses, federal facilities, local governments and tribes with tools to help meet environmental regulatory requirements.

Underground Storage Tanks. EPA enforces requirements under Subtitle I of the Resource Conservation and Recovery Act. These requirements focus on preventing, detecting, and cleaning up releases. These provisions are enforced by EPA and by states that are authorized to operate their own program in lieu of the federal program.

Lead-based paint. EPA enforces requirements under the Residential Lead-Based Paint Hazard Reduction Act. Contractors and construction professionals who work in pre-1978 housing or child-occupied facilities must follow lead-safe work practice standards to reduce lead exposure. This includes providing owners, tenants, and child care facilities with lead-based paint information; and notifying them about the presence of lead. Owners and landlords of pre-1978 residential housing must give tenants a lead-based paint warning pamphlet and notify the tenants of known lead-based paint in the housing. Sellers are subject to similar requirements.

Asbestos. EPA enforces regulations under the Asbestos Hazard Emergency Response Act (AHERA) on how to respond to asbestos in schools. EPA also enforces worker protection standards for certain state and local government employees who are not protected by the asbestos standards of the Occupational Safety and Health Act.

Accidental Releases. EPA enforces requirements under Section 112(r) of the Clean Air Act to prevent chemical accidents and releases. Owners and operators of sources producing, processing and storing extremely hazardous substances must identify hazards associated with an accidental release, design and maintain a safe facility, prepare a Risk Management Plan (RMP) and minimize consequences of accidental releases that occur. EPA conducts inspections and reviews facility RMPs to verify compliance and ensure the quality of the overall preparedness, prevention and response.

Chemical Enforcement

Pesticides. EPA enforces requirements under the Federal Insecticide Fungicide and Rodenticide Act (FIFRA) that govern the distribution, sale and use of pesticides. EPA takes enforcement actions to address the distribution or sale of unregistered pesticides, registered pesticides whose composition differs from that submitted at registration, and registered pesticides that are misbranded or adulterated. EPA may also stop the sale or seize pesticide products which do not meet FIFRA requirements. One focus of EPA's FIFRA enforcement program is to ensure pesticides entering the United States meet FIFRA requirements.

EPA and the states verify FIFRA compliance through a comprehensive FIFRA compliance monitoring program which includes inspecting facilities, reviewing records and taking enforcement action where necessary. The FIFRA compliance assistance program provides businesses, federal facilities, local governments and tribes with tools to help meet environmental regulatory requirements.

Toxic Chemicals. EPA enforces requirements under the Toxic Substances Control Act (TSCA) which regulates the introduction of new or already existing chemicals. TSCA requirements for chemical manufacturers or importers include reporting, record-keeping and testing of the chemical substances.

EPA and the states verify TSCA compliance through a comprehensive TSCA compliance monitoring program which includes inspecting facilities, reviewing records and taking enforcement action where necessary. The TSCA compliance assistance program provides businesses, federal facilities, local governments and tribes with tools to help meet environmental regulatory requirements.

PCBs. EPA enforces regulations under TSCA. TSCA prohibits the manufacture of polychlorinated biphenyls (commonly known as PCBs), controls the phase-out of their existing uses, and sees to their safe disposal.

Emergency Planning and Community Right to Know (EPCRA)

The Emergency Planning and Community Right-to-Know Act (EPCRA) was passed by Congress in response to concerns regarding the environmental and safety hazards posed by the storage and handling of toxic chemicals. These concerns were triggered by the 1984 disaster in Bhopal, India, caused by an accidental release of methyl isocyanate. The release killed or severely injured more than 2000 people. To reduce the likelihood of such a disaster in the United States, Congress imposed requirements for federal, state and local governments, tribes, and industry. These requirements covered emergency planning and "Community Right-to-Know" reporting on hazardous and toxic chemicals. The Community Right-to-Know provisions help increase the public's knowledge and access to information on chemicals at individual facilities, their uses, and releases into the environment. States and communities, working with facilities, can use the information to improve chemical safety and protect public health and the environment.

EPA enforces requirements under EPCRA to ensure that facilities are prepared for chemical emergencies and report any releases of hazardous and toxic chemicals. EPCRA requires that citizens be informed of toxic chemical releases in their area. Industrial facilities must annually report releases and transfers of certain toxic chemicals. This information is publicly available in the Toxics Release Inventory (TRI) database. EPA and the states verify EPCRA compliance through a comprehensive EPCRA compliance monitoring program which includes inspecting facilities, reviewing records and taking enforcement action where necessary. The EPCRA compliance assistance program provides businesses, federal facilities, local governments and tribes with tools to help meet environmental compliance.

Cleanup Enforcement

EPA's cleanup enforcement program protects human health and the environment by getting those responsible for a hazardous waste site to either clean up or reimburse EPA for its cleanup. EPA uses a number of cleanup authorities independently and in combination to address specific cleanup situations.

- **Superfund.** Under the Comprehensive Environmental Response, Compensation, and Liability Act (CERCLA, commonly known as Superfund), EPA finds the companies or people responsible for contamination at a site, and negotiates with them to clean up the site themselves or to pay for another party to do the cleanup.

- **Corrective Action.** When solid or hazardous waste is not properly managed and contamination results at facilities regulated by the Resource Conservation and Recovery Act (RCRA), EPA and the states may step in to oversee the cleanup.

- **Leaking Underground Storage Tanks.** RCRA provides EPA with several enforcement authorities to ensure that releases from leaking underground storage tanks (USTs) are cleaned up and owners/operators comply with other RCRA requirements.

- **Brownfields and Land Revitalization.** The sustainable reuse of previously contaminated property is an important goal of EPA's hazardous waste cleanup programs.

OIL POLLUTION ACT (OPA)

Summary

The Oil Pollution Act (OPA) was signed into law in August 1990, largely in response to rising public concern following the Exxon Valdez incident. The OPA improved the nation's ability to prevent and respond to oil spills by establishing provisions that expand the federal government's ability, and provide the money and resources necessary, to respond to oil spills. The OPA also created the national Oil Spill Liability Trust Fund, which is available to provide up to one billion dollars per spill incident.

Originally published in 1973 under the authority of §311 of the Clean Water Act, the Oil Pollution Prevention regulation sets forth requirements for prevention of, preparedness for, and response to oil discharges at specific non-transportation-related facilities. To prevent oil from reaching navigable waters and adjoining shorelines, and to contain discharges of oil, the regulation requires these facilities to develop and implement Spill Prevention, Control, and Countermeasure (SPCC) Plans and establishes procedures, methods, and equipment requirements (Subparts A, B, and C).

In 1990, the Oil Pollution Act (OPA) amended the Clean Water Act to require some oil storage facilities to prepare Facility Response Plans (FRP). On July 1, 1994, EPA finalized the revisions that direct facility owners or operators to prepare and submit plans for responding to a worst-case discharge of oil (Subpart D).

Federal Facility Responsibilities under the Oil Pollution Prevention Program

- Developing and updating the facility's oil spill emergency response plans

- Maintaining required records/documentation

- Testing emergency response equipment

- Periodically performing mock spill response drills

- Notifying federal, state, and local agencies in case of an incident

- Mitigating all spills and discharges

- Ensuring employees have required training

Application of Oil Pollution Prevention Regulation to Federal Facilities

OPA requires EPA to amend the National Oil and Hazardous Substances Pollution Contingency Plan (NCP) to enhance and expand procedures for oil spill response. In addition, OPA requires certain facilities to develop response plans for responding to worst-case discharges of oil and hazardous substances. Federal facility activities subject to OPA requirements include:

- Storing or handling petroleum, fuel oil, sludge oil, and oil mixed with waste;
- Transferring oil by using motor vehicles or rolling stocks;
- Supporting maritime vessel activities or other water-related activities where fuels are used.

EPA Enforcement

EPA's enforcement authorities for OPA violations reside in CWA §311(e) and §311(c). Typically, EPA will negotiate a compliance agreement with a federal agency in violation of OPA. The typical compliance agreement contains several provisions including schedules for achieving compliance and dispute resolution.

Criminal Enforcement

Individual: Employees of federal facilities may have criminal sanctions brought against them for violations of OPA. Criminal fines may be imposed for violations of OPA under CWA §309.

    State Enforcement: Section 1019 of OPA authorizes states to enforce, on the navigable waters of the state, the requirements for evidence of financial responsibility under OPA §1016.

    Tribal Enforcement: OPA contains no Tribal enforcement provisions.

    Citizen Enforcement: OPA contains no citizen enforcement provisions.

EPA OPA Regulations

OPA regulations are set forth in:

- 40 CFR Parts 110, 112, and 300 subparts C, D, E
- 49 CFR Part 194
- 33 CFR Part 154

THE PHASE-OUT OF OZONE-DEPLETING SUBSTANCES

EPA regulations issued under Sections 601–607 of the Clean Air Act phase out the production and import of ozone-depleting substances (ODS), consistent with the schedules developed under the Montreal Protocol. The U.S. phase-out has operated by reducing in stages the amount of ODS that may be legally produced or imported into the U.S. The Parties to the Montreal Protocol have changed the phase-out schedule over time, through adjustments and amendments, and EPA has also accelerated the phase-out under its Clean Air Act authority. As the phase-down of virgin ODS continues, ODS uses will increasingly resort to reclaimed material or alternatives.

In the United States, ozone-depleting substances are regulated as Class I or Class II controlled substances. Class I substances have a higher ozone-depleting potential and have been completely phased out in the U.S., except for exemptions allowed under the Montreal Protocol. Class II substances are hydrochlorofluoro-carbons (HCFCs), which are transitional substitutes for many Class I substances and are being phased out now.

EPA has issued regulations under Section 608 of the Clean Air Act to minimize the emission of refrigerants by maximizing the recovery and recycling of such substances during the service, repair, or disposal of refrigeration and air-conditioning equipment (i.e., appliances).

Note: The handling and recycling of refrigerants used in motor vehicle air-conditioning systems are governed under Section 609 of the Clean Air Act.

Refrigerant Leak Repair

The leak repair requirements, promulgated under Section 608 of the Clean Air Act, require that when an owner or operator of an appliance that normally contains a refrigerant charge of more than 50 pounds discovers that refrigerant is leaking at a rate that would exceed the applicable trigger rate during a 12-month period, the owner or operator must take corrective action.

UNIVERSAL WASTES

EPA's universal waste regulations streamline hazardous waste management standards for federally designated "universal wastes," which include:

- batteries

- pesticides

- mercury-containing equipment and

- bulbs (lamps)

The regulations govern the collection and management of these widely generated wastes, thus facilitating environmentally sound collection and proper recycling or treatment.

These regulations also ease the regulatory burden on retail stores and others that wish to collect these wastes and encourage the development of municipal and commercial programs to reduce the quantity of these wastes going to municipal solid waste landfills or combustors. In addition, the regulations also ensure that the wastes subject to this system will go to appropriate treatment or recycling facilities pursuant to the full hazardous waste regulatory controls.

The federal universal waste regulations are set forth in 40 CFR part 273. States can modify the universal waste rule and add additional universal waste(s) in individual state regulations so check with your state for the exact regulations that apply.

References

ASQ/ANSI/ISO 14001:2015. *Environmental management systems—Requirements with guidance for use.* September 2015.

Haddadin, Jim. 2015. "Marlborough Lab Company Hit with $100K Environmental Fine." *MetroWest Daily News*, December 4. http://www.metrowestdailynews. com/article/20151204/NEWS/151207721. Reprinted by permission of the author.

Occupational Safety and Health Administration—OSHA 3084 1998, Federal Register.

Index

Note: Page numbers followed by *f* or *t* refer to figures or tables, respectively.

About the Author

Milton Dentch was born and raised in the Worcester, Massachusetts, area. He has a BS in mechanical engineering from Worcester Polytechnic Institute and an MS in quality management systems from the National Graduate School of Quality Management (NGS). After college, he worked as an engineer in the paper industry for 5 years, then he worked as an engineer and manager at the Polaroid Corporation in Waltham, Massachusetts, for 27 years. He was plant manager for the Custom Coating and Laminating plant in Worcester for the Furon Corporation. Milt has over 40 years' experience in a wide variety of industries, including pulp and paper, chemical, plastic and rubber processing, battery manufacturing, converting, electronics assembly, and machine building.

Milt currently provides consulting, training, and auditing related to the International Organization for Standardization requirements for quality, environmental, and safety management systems. He has conducted over 500 audits worldwide for large and small companies. His clients have been as diverse as a floating oil rig in the Gulf of Mexico to an electronics manufacturer in the Ukraine with 4000 employees. Milt is an Exemplar Global qualified Lead Auditor for Quality and Environmental Management Systems and a Registrar approved OHSAS 18001 Lead Auditor.

In 2012, Milt wrote *Fall of an Icon—Polaroid after Edwin H. Land* (RiverHaven Books), an insider's history of the Polaroid Corporation.

www.ingramcontent.com/pod-product-compliance
Lightning Source LLC
Chambersburg PA
CBHW080758300326
41914CB00055B/937

* 9 7 8 0 8 7 3 8 9 9 2 9 1 *